D0765601

STATISTICAL METHODS IN GEOLOGY

TITLES OF RELATED INTEREST

Rutley's elements of mineralogy, 26th edn
H. H. Read

Petrology of the igneous rocks, 13th edn
F. H. Hatch, A. K. Wells & M. K. Wells

Metamorphism and metamorphic belts
A. Miyashiro

Metamorphic processes
R. H. Vernon

Physical processes of sedimentation
J. R. L. Allen

The formation of soil material
T. R. Paton

Petrology of the metamorphic rocks
R. Mason

The interpretation of igneous rocks
K. G. Cox, J. D. Bell & R. J. Pankhurst

Petrology of the sedimentary rocks, 6th edn
J. T. Greensmith

Invertebrate palaeontology and evolution
E. N. K. Clarkson

A dynamic stratigraphy of the British Isles
R. Anderton, P. H. Bridges, M. R. Leeder & B. W. Sellwood

Microfossils
M. D. Brasier

The inaccessible Earth
G. C. Brown & A. E. Mussett

Tectonic processes
D. Weyman

Geomorphological techniques
A. S. Goudie (ed.)

Terrain analysis and remote sensing
J. R. G. Townshend (ed.)

British rivers
J. Lewin (ed.)

Soils and landforms
A. J. Gerrard

Sedimentology: process and product
M. R. Leeder

Introduction to small-scale geological structures
G. Wilson

Metamorphic geology
C. Gillen

The poetry of geology
R. M. Hazen

Sedimentary structures
J. D. Collinson & D. B. Thompson

Komatiites
N. T. Arndt & E. G. Nisbet (eds)

Aspects of micropalaeontology
F. T. Banner & A. R. Lord (eds)

Geomorphological field manual
R. Dackombe & V. Gardiner

Geology & Man
Janet Watson

Field mapping for geology students
F. Ahmed & D. Almond

STATISTICAL METHODS IN GEOLOGY

for field and lab decisions

R. F. Cheeney
Grant Institute of Geology,
University of Edinburgh

London
GEORGE ALLEN & UNWIN
Boston Sydney

**George Allen & Unwin (Publishers) Ltd,
40 Museum Street, London WC1A 1LU, UK**

George Allen & Unwin (Publishers) Ltd,
Park Lane, Hemel Hempstead, Herts HP2 4TE, UK

Allen & Unwin Inc.,
9 Winchester Terrace, Winchester, Mass. 01890, USA

George Allen & Unwin Australia Pty Ltd,
8 Napier Street, North Sydney, NSW 2060, Australia

First published in 1983

British Library Cataloguing in Publication Data

Cheeney, R. F.
Statistical methods in geology for field and lab decisions.
1. Geology – Statistical methods
I. Title
519.5′024553 QE33.2.M3
ISBN 0–04–550029–0
ISBN 0–04–550030–4 Pbk

Library of Congress Cataloging in Publication Data

Cheeney, R. F.
Statistical methods in geology for field and lab decisions.
Bibliography: p.
Includes index.
1. Geology – Statistical methods. I. Title
QE33.2.M3C48 1983 550′.72 83–7647
ISBN 0–04–550029–0
ISBN 0–04–550030–4 (pbk)

Artwork drawn by Oxford Illustrators Ltd

Set in 10 on 12 point Times by
Mathematical Composition Setters Ltd, Salisbury, Wilts
and printed in Great Britain by
Richard Clay (The Chaucer Press) Ltd.
Bungay, Suffolk.

Preface

Here's a book on statistics, written by a field geologist, not a statistician. Having thus damned the volume in the first sentence, let me try now to resurrect a semblance of utility. Geology is dominantly observational, and progress on both pure and applied fronts depends on skilled documentation. Such careful records support consolidation of understanding and experienced insight, and provide also for creative intuition. Theories proliferate and data accumulate, and to ensure effort is channelled in appropriate directions, interpretations must be sound. We need to make decisions; the nearer to source observations, the better. When we stand looking at a rock exposure, drill core or whatever, we see possibly 5% of what is there; of what we see, perhaps we understand 5% again. We must sharpen our observational skills and be confident in our interpretational hypotheses. This is where the statistics arrives – as a data-handling and decision-making tool. If we can decide promptly that a given interpretation is applicable, then we have made a finite advance upon which further work can be based. Advance depends on the acquisition of data, which is often costly. Can we improve efficiency, as well as decision making, by tailoring the amount of data to the needs in hand, or vice versa, can we extract more information from the data already in hand? The answer to both questions is 'very likely'. Usually we require simple answers to simple questions but the routes can be devilishly tortuous.

So, about this book. What kind of mathematics do you need? Not much more than the kind of school-level algebra and trigonometry you use in map reading and basic survey work, without which you probably shouldn't be out in the bush anyway. If your map-reading skills include handling spherical projections, so much the better, because in Chapter 9 (on three-dimensional orientation data) I speed up a bit. There's some really basic matrix algebra too, but I have included a glossary that should cover all that's needed.

Do tackle the examples and exercises, few as they are. They are all authentic to the extent that I have retained Imperial units in some because that's the way the original measurements were made. They also have a good Scottish kailyard look to them: all collected within easy reach of Edinburgh. We like to think we have classical geology here, engendered by the work of James Hutton, but I don't think it's particularly unusual, certainly not unique. The point I must emphasise, *most strongly*, is that your own kailyard gives just as much opportunity for exercise, even before you reach the end of Chapter 1. So I hope I despatch the parochial impression and encourage you to search out your own examples – the sooner, the better. The methods I employ can be used anywhere.

Lastly, who do I think you are, and how should you use this book? You may be a student, with a year of geology behind you, or a practising industrial or academic geologist. Although study of statistical methods is creeping into geology degree courses, it's by no means universal – so perhaps this book will help to fill a small gap. You use it partly for learning, partly for reference but mostly as a handbook. The idea is for you to carry it with you, so the more dog-eared it gets, the happier I shall be. When you have a problem, pull out this book, a pencil, a notebook, maybe a calculator, and decide on the spot. Do write if you find any mistakes (or should I say *when* you find them). I take full responsibility to the extent that I have recalculated all the tables, except two small ones as referenced. All the methods have been programmed for the Hewlett-Packard HP41C machine although I don't give them here because you will all have different calculating aids at hand. I await the day they build the HP41C into a hammer handle!

R. F. Cheeney
Penicuik, Scotland
November 1982

Acknowledgements

Writing a book is a team effort. My team has been small but effective. Pride of place must go to Mrs Patricia Stewart who typed the entire first version of the manuscript with great patience, although sometimes reduced to muttering in her native Erse at the fiddly mathematical expressions. My reviewers worked nearly as hard, correcting, advising and stimulating. Dr Roger Mason of the Department of Geology, University College London, helped tremendously to clear the verbal jungle surrounding the earlier version. Dr Nigel Woodcock of the Department of Earth Sciences, University of Cambridge, was especially encouraging with respect to orientation statistics. Dr Lee Belbin of the Institute of Biological Resources, CSIRO, Canberra, provided a wealth of advice, and I am particularly grateful for his remarks on parameters, degrees of freedom, and numerical/computational procedures. Dr Dan Merriam, writing from Wichita State University, Kansas, gave me deep insight into differences of usage of English between Scotland and North America. I fear I have not acted on a number of his suggestions and apologise to North American readers for retaining 'further', 'colour', etc., despite his reservations. He provided also a wealth of detailed observation, too voluminous to summarise here. Lastly, Dr Iaakov Karscz of the State Geological Survey of Israel, a valued associate of long standing, offered characteristically thoughtful remarks: I feel my response to many of his suggestions is somehow inadequate.

The task of review complete, the second round of typing was accomplished in days by Mrs Marcia Wright (Introduction and Chs 1 & 2), Mrs Patricia Scrutton (Chs 3, 4 & 5), Mrs Ulla Hipkin (Chs 6, 7 & 8) and Mrs Margaret Swift (Chs 9 & 10).

Roger Jones, Geoffrey Palmer, Elizabeth Royall and colleagues at Messrs George Allen & Unwin have been instrumental in guiding this project to fruition.

Contents

List of tables

List of critical values, useful expressions, etc.

A note on evaluation of expressions

The order of priorities in evaluating some of the more complicated expressions in this book is important, and is as follows:

(1) calculate the contents of the innermost brackets first, following which:
(2) raise to powers, take square roots ($\sqrt{}$), logarithms, sines, cosines and tangents, then:
(3) multiply ($\times$) or divide (/) from left to right, then:
(4) add or subtract from left to right.

Note that operations of equal priority are executed from left to right. If you use an electronic calculator, be sure you are familiar with the inbuilt priorities that your machine gives to the basic arithmetical operations, because some give priority to multiplication and division over addition and subtraction whereas others do not.

Example In X-ray diffraction, we make frequent use of the Bragg-Law equation: $\lambda = 2d \sin\theta$, rearranged so as to calculate the lattice-plane spacing:

$$d = \lambda/(2 \sin\theta)$$

With $\lambda = 1.54$ angstrom units and $\theta = 35.2°$, and a calculator set to display results to three decimal places, evaluate the bracketed part of the expression first:

$$2 \sin\theta = 2 \sin 35.2° = 2 \times 0.576 = 1.153$$

Then perform the division:

$$d = 1.54/1.153 = 1.336 \text{ angstroms}$$

Introduction

There are four things that I want to do in this brief introduction: first, to discuss various meanings that are attached to the word *statistics* and to suggest some kinds of applications that arise with each: secondly, to describe the organisation of the book, especially some of its less usual features, and to demonstrate the logical connection between chapters: thirdly, to outline the settings in which the methods may be applicable: and fourthly, to suggest the rôles that may be played by calculators and computers.

'Statistics' signifies different things to different people and certainly prompts a range of responses; from complete rejection, through suspicion, then guarded acceptance, to unbridled admiration. Similarly, applications range in character from criminal abuse to arcane sophistication. Let me try to plot a middle course (and certainly arguable) by beginning with the suggestion that 'statistics' and statistical methods fall into one of four categories, as outlined below.

(a) First, there is the lay person's idea of 'statistics', namely, data (not always numerical) gathered for its own sake and presented in either graphical or tabular form as a *complete description* of a collection of objects. Examples include a list of all the fossils extracted from a locality, and a stereographic projection showing the three-dimensional orientations of all the dykes in a study area. We accumulate such descriptions as the basis for any exercise in the observational sciences and they may constitute the end product if the objectives are so framed.

(b) Numerical parameters calculated from the original data may be used *so as to summarise* such a collection. Such descriptive parameters are generally calculated using simple mathematical expressions. Examples include the *proportions* of different fossil species in the collection as a whole, and the *average* orientation of dykes in a study area. Other examples of such parameters include the widely used **mean** (a measure of the average value of a variable quantity) and **standard deviation** (a measure of the dispersion of the variable quantity about its mean value). The advantages of calculating such parameters are clear in that they allow whole tables of numbers to be replaced by one or two such descriptive parameters. Although much detail is lost, the gain in brevity can be very useful.

(c) Perhaps one of the more powerful applications of statistics lies in its use *to test hypotheses*, which is given emphasis in this book. To develop the previous examples: Is the proportion of brachiopods in the fossilised community at this locality *significantly different* from the proportion at some other specific locality? Is the average orientation of dykes in this area *significantly different* from that in the area to the north? As shown in Chapter 2, making a claim that values of a descriptive parameter (or whatever) are significantly different from one place to another simplifies to calculating the value of the probability that they are the same, and then arguing that this probability is so small that it can be safely ignored. Successful application of hypothesis-testing methods allows subjective phrases such as '... it is possible that ...' to be replaced by '... it is probable that ...' with the additional bonus that this probability may be quantified to make the whole statement quite objective.

(d) Lastly, when quoting a value for a descriptive parameter or for a number arising from any experimental procedure, such as a chemical analysis, it is very useful to be able to make some statement about our confidence in the quoted figure. Whether such a figure is a descriptive parameter or an experimental result, we can never pretend that it will coincide exactly with the 'true' value: it must always remain an estimate. How close it is likely to be can often be calculated. To continue with our previous examples: the 'true' proportion of brachiopods in this fossilised community lies between such and such lower and upper limits, with probability so and so; the 'true' average orientation of these dykes lies within this specified area of the stereogram, with probability so and so.

Now let me describe the organisation and scope of the book, as summarised in Figure 0.1. I make no apology for the fact that I concentrate on methods, so let me draw an imperfect analogy. I have an old car and a tool-kit. The objective is to use the tool-kit to keep the old car going. The correct functioning of the old car represents advancing knowledge, the tool-kit the means of keeping it going. The car consists of several systems: the engine –clutch–gearbox, the brake system, the electrical system, the

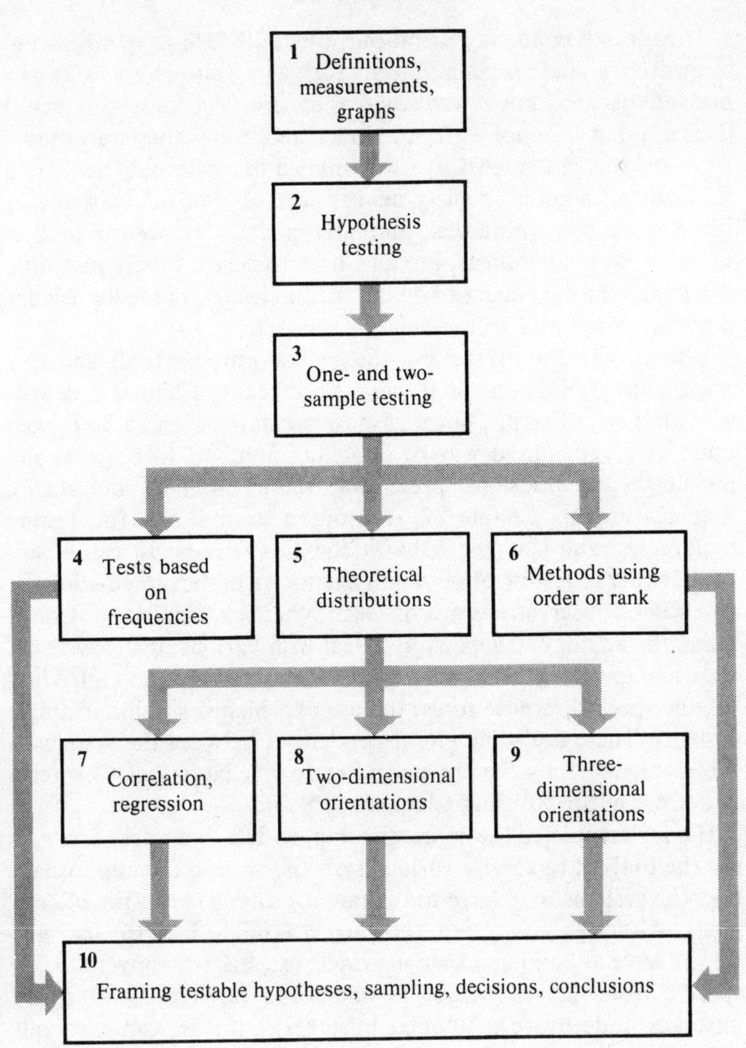

Figure 0.1 Logical, though not quite watertight, pathways through the book. The numbers are those of the corresponding chapters.

wheels and tyres, etc. These systems represent different, but interdependent, branches of my advancing knowledge. From time to time, one of these systems will break down. Sometimes I fit a new component, or replace a whole system. In geology, one of my cherished hypotheses had a weak, but replaceable link, or was completely unfounded. Either way, the tool-kit is in the boot and I can bring it out whenever I need it.

The tool-kit contains general-purpose tools such as spanners and screwdrivers, and specialised tools such as a voltmeter and an air-pressure gauge. I know what these tools are, and something about their use, but I do not worry too much about how they were made. I can see that the screwdriver was shaped out of metal, and could fabricate a makeshift replacement if need be, but the voltmeter is just a black box with a dial. Similarly in statistics, we can look at some of the simpler ideas, but may have to accept others with little more than the equivalent of an instruction leaflet, in the knowledge that they work and will certainly be useful.

Chapters 1, 2 and 3 set out the general-purpose tools and give some general hints on using them. Specifically, Chapter 1 defines some important terms, suggests how measurement can be looked upon as a generalised way of attaching numbers to objects, and introduces methods of presenting these numbers in concise graphical form. Chapter 2 sets out a framework for testing hypotheses, and Chapter 3 shows how this framework can be applied, either to testing observation against theoretical prediction, or two sets of observations against each other for 'similarity'. Continuing the analogy, Chapters 4–9 deal with various of the systems, showing how the general-purpose tools may be used, and introducing the special-purpose tools, the use of which is specific to single systems. There are some interdependencies between the systems – these are shown by the arrows in Figure 0.1. Note that Chapter 5 is a prerequisite to using Chapters 7, 8 and 9.

Having completed the apprenticeship, and having learned how to use the tool-kit to service various systems, we can then appreciate how the systems may have to operate together to keep the old car going. Although we are now thoroughly familiar with the tool-kit, we still have to build up experience in its use, discover how to tackle specific tasks in an orderly manner, and how to avoid costly mistakes. Indeed, we shall make mistakes, but if we can carry out careful 'post-mortem' analyses, we shall benefit by these experiences. Chapter 10 reviews the methods, and draws some conclusions about our approach to the statistical aspects of our projects. Some reviewers have argued that Chapter 10 should be Chapter 1. There are good reasons either way, but I have finally chosen to set out the tool-kit first and describe the old car last. One advantage of this is that we can pick out various tools along the way and try them out on inconsequential odd jobs, the outcome of which is not critical, but the experience gained is useful in the learning and appreciation process.

One final note concerning a detail of the organisation of the book: many of the methods used require that we consult formulae or tables of values of some variable. Rather than segregate these formulae and tables into an appendix, I have placed them in the text near to their points of introduction. A complete list is given at the beginning of the book.

Next, why and where are these statistical methods of use to geologists? Much geological field and laboratory work consists of gathering data which may simply be written down or displayed on a map. Initially, these data may not be highly organised because their purpose is solely descriptive. But we should soon begin to speculate about aspects of the data so as to develop an interpretation that will rationalise seemingly unrelated strands. This should help to evolve an understanding that we can pass on to others in a more digestible form. At the lowest level of interpretation, our hypotheses are likely to be unrefined – little more than general impressions – and rarely statistically testable in their original form. However, we should try to be objective near the outset, so we need a selection of methods that will allow us to explore some of the features of our data; features whose presence may be shadowy or even totally hidden. If the methods are to be exploratory and practical, then we need just sufficient statistical theory to understand how to frame statistically testable hypotheses. Testing these hypotheses in the immediacy of the field or laboratory allows us to make the objective statement, or draw the sound conclusion, that takes us to the next step in our project. The nearer to the data source that we are able to do this, the better, because in concentrating our attention in this way, we are really adding to our powers of perception and sharpening our objectivity. Furthermore (and very important), we may find that we can make economies in both the amount and type of data that we need to gather, and we have to remember that data acquisition is perhaps the most costly part of the whole process.

The mathematics must be kept simple because the methods are exploratory, and applicable near to the data source. The book is likely to serve as a handy source of reference and so it must be concise, with methods that minimise the labour of calculation consistent with the objectives in hand. There must be some calculation, though, and a basic scientific calculator is a useful asset that will cover 90% of applications here. The calculator should have the four basic arithmetic functions (addition, subtraction, multiplication, division); the trigonometrical functions and their inverses;

square root and reciprocal; natural logarithms (i.e. to base 'e'); an addressable memory (to hold a constant or to accumulate intermediate results); and so-called 'scientific' (or mantissa and exponent) display, because the numbers generated will range from very small to very large. A five-figure display should be adequate, but internally, the numerical precision should be better than this (see Appendix C.7). However, for orientation data, especially in three dimensions, a key-programmable calculator, with ten or more addressable memories, can be helpful for *in-situ* analyses of joint surfaces, fold orientation, sedimentary direction, etc. Indeed, such has been the pace of technological advance during the time my reviewers and I have laboured over this manuscript that it is now possible to pack *all* of the tables and formulae in this book, and to manipulate them automatically in a pocket-sized programmable machine with continuous memory. By far the most labour intensive part of any project is now so clearly the data acquisition stage.

I have not made reference to the types of computers found in the office or laboratory because I do not want to assume that my readers will have access to them. Packages of statistical programmes are often mounted on such machines, but they commonly use methods that are more refined than those presented here, or whose data sets have a more complicated structure. I have looked upon such methods as being beyond the scope of this book, save that Chapter 7 may provide an avenue of introduction, and I would refer the interested reader to the Bibliography.

1 Words, numbers and pictures

This chapter begins with definitions of three important terms: specimen, sample and population. It goes on to discuss measurement in the most general terms, as a procedure for attaching numbers to objects or events. Lastly, some methods are introduced whereby these numbers can be arranged and presented in a more digestible pictorial or graphical form. Some of the graphs are suited to immediate statistical testing or can be used to estimate descriptive parameters directly.

1.1 Three definitions

Strict discipline in the usage of terms is crucial to avoid confusion.

specimen (or **individual** or **event**) The single basic unit or object of study: e.g. one pebble from a beach, one rock specimen from a formation, one fossil from a bed, one stratigraphical event (perhaps incoming of clay) in a succession, etc.

sample A finite number of specimens selected according to some plan (the 'sampling plan') to provide the basic information for the study or 'experiment' (the latter used in a very general sense): e.g. a bucketful of pebbles from the beach, a sackful of rock specimens from the formation, a collection of fossils from the bed, a stratigraphical record along some coastal cliffs. (Please note and remember the difference between specimen and sample!)

population The total number of all possible specimens in the study: e.g. all the pebbles on the beach, all the rock comprising the formation, all the fossils that ever occurred in the bed, the world-wide stratigraphy of the Carboniferous Period. Further concepts attached to 'population' are developed in Section 10.3.2.

1.2 Measurement

Fundamental to any study that involves a statistical approach is the laying down of a sampling plan which, in brief, comprises three steps: a) a description of the 'target' population; (b) a set of rules for drawing a sample from that population; (c) an account of the types of observations to be made on each specimen. Here, I deal only with the last of these; the first two are treated in detail in Chapter 10.

Traditional geological training introduces the idea that rocks and fossils can be classified according to some scheme or other. Further, at junior school, we learn that measuring the length of, for example, a fossil means laying a ruler alongside the fossil and noting its length in centimetres and fractions thereof. These are only two ideas under a more general umbrella of 'measurements'. If we can think of **measurement** as a means of distinguishing between (or comparing) specimens, then we can measure by attaching a number (or a name) to each specimen in a sample. The act of using these numbers or names implies that we have a **scale of measurement** at our disposal. Two broad groupings of such scales are **linear** (lengths, weights, densities) and **angular** (orientations in two or three dimensions). Leaving the latter until Section 1.2.2, consider linear scales of measurement, amongst which we can distinguish three important categories:

(a) **Nominal** scale with 'defining arithmetic' of 'equivalence'. An example is a classification in which a fossil is allotted to a given species on the basis of certain of its morphological features. The species may be identified either by name (*Monograptus priodon*) or by number (graptolite type 6). Allocating a number in this way serves only as a convenient means of labelling, and when we draw a comparison between a graptolite type 6 and one of type 3, most of what we can say adds up to their being different. More succinctly, we may say that nominal-scale measurements refer to a lack of ordered differences. Pedantic as it may seem at this stage, the only valid arithmetic operations on nominal scale numbers are 'equality' and its inverse, 'inequality'.This is a philosophical point but is occasionally of practical importance. Related to the nominal scale in some ways is the concept of positive and negative 'attributes' that I introduce in Section 4.1.

(b) **Ordinal** scale with defining arithmetic of equivalence or 'greater than'. Density of coloration (reflecting haematite content), as shown by specimens of brown sandstones, might serve as an example here. Although it might be possible to measure density of coloration photometrically on an absolute scale, it is certainly easier to arrange such specimens in a line in order of increasing colour density. This accomplished, the individuals can then be numbered 1, 2, 3, etc., starting with the palest, and we know immediately that specimen 6 is darker than specimen 3, but not necessarily twice as dark. The valid arithmetic operations on ordinal scale numbers are equality and 'greater than' and their inverses.

(c) **Ratio** scale with defining arithmetic of that of the ordinal scale plus the known and fixed ratio of any two scale values. Ratio scales are abundant and familiar and correspond to the bulk of the measurements of the layperson. Examples include mass, length, time and the defining arithmetic above means, for example, that the mass of an object in English pounds divided by the mass of the same object in kilograms is a constant termed the conversion factor. Valid arithmetic operations on numbers belonging to this scale include not only equality and 'greater than', but also the more customary operations of addition and multiplication with their inverses, subtraction and division.

1.2.1 Continuous and discontinuous scales

Before leaving linear measurement, the concept of **continuous** versus **discontinuous** scales should be mentioned. Clearly, the nominal scale is a discontinuous scale because it has only a finite number of discrete points which may be attached to specimens. Hence, we may have graptolite types 3 and 6, but not types 4.5 or 3.2. Numbers on the ordinal scale can take only positive, integer (i.e. whole number) values so that they also form a discontinuous set. However, in theory at least, ratio measurements can be made to an arbitrarily high level of precision, with only a practical limit to the number of decimal places that may be quoted, and so constitute a continuous scale of measurement.

For discussion. Prismatic columns developed by jointing of basalt can be measured by counting the number of their polygonal faces. What is the scale of measurement in the above scheme in this case? What is the total range of the numbers on this scale? In particular, is the scale continuous or discontinuous and how may an 'average' be defined?

1.2.2 Orientations

Unfortunately, no commonly accepted usage of terms is current in treatments of orientation (or 'directional' or 'angular') data. In the next two sections, I define *directions* and *axes* and treat these as being two different types of *orientations*. My orientations have in common that they are principally specified by angles and I note that measurements of orientations, in both two and three dimensions, have some features in common with those on linear scales, but in other respects are totally, sometimes deceptively, different. Thus, orientation measurements are clearly *continuous*: however, the origin (i.e. the zero from which we measure) may be arbitrary in that we may choose to measure azimuth either from geographical north or magnetic north (or wherever else may be convenient) and inclination either from the horizontal or vertical. They may also have the property of cyclicity. Thus azimuths repeat, with a period of 360° (i.e. once round the compass), whereas inclinations do not in general, and have more restricted ranges (usually ± 90°). A property of cyclicity is sometimes useful: time is cyclical (daily, weekly, yearly) as are certain sedimentary deposits, volcanic phenomena, etc. Consequently, linear data which have an underlying cyclicity can be transformed into angles and analysed by some of the methods of the statistics of orientation data. Arithmetic manipulation of angles, especially in two dimensions, as round the compass, holds some hazards for the unwary. Thus, in certain situations, we may add angles, or double them, but beware of calculating 'straightforward' averages. The average of 2° and 358° is seemingly 180° but this would be nonsense if the study were of sedimentary directions. This is because orientation measurements have many of the properties of vectors (i.e. line elements that have orientation, 'sense' or 'polarity', and magnitude) and so must be manipulated accordingly. Consequently we require two more definitions.

direction (directional data) A line specifying an orientation (two or three dimensions) associated with which is a 'sense' (which may be symbolised by an arrowhead at one end of the line). Examples are palaeomagnetic directions, directions of sedimentation or sedimentary linear structures, normals to bedding in the direction of younging, etc. With the additional property of 'magnitude', the 'direction' becomes a 'vector' (e.g. wind velocity) but vectors are rarely needed in this book.

axis (axial data) A line that specifies an orientation (again, in two or three dimensions) but no more, in particular there is no 'sense'. Examples include the normal to a dyke, the length of an acicular crystal and, unfortunately, the axial 'direction' of a fold. The last example is but one of many that will illustrate the current confusion of usage of the terms ‘axis’, ‘direction’ and ‘vector’.

1.2.3 Choice of appropriate scale of measurement

An integral part of the design of the sampling plan (reviewed in Ch. 10) is the choice of an appropriate scale of measurement. In certain situations, common in practice, one may choose between two, or even three, scales of measurement (‘nominal’, ‘ordinal’ or ‘ratio’). One important factor for early consideration is that of the labour involved in taking measurements – data acquisition *is* expensive. Thus it could be that some hypothesis relates to measurements made on the grain size of rock specimens from a formation. Grain size can be measured on a ratio scale (with a micrometer), on an ordinal scale (by laying the rock specimens out in a line in order of increasing grain size), or on a nominal scale (by classifying each specimen as ‘fine-grained’ or ‘medium-grained’ or ‘coarse-grained’). Clearly, the ratio-scale measurements are slowest, and the nominal-scale measurements fastest. The hypothesis under test may dictate the choice but, on the contrary,it is very likely that in many practical situations the framing of the hypothesis may be influenced by the ease or otherwise of measurement. If the outcome is tolerably independent of the choice of method, or if we are carrying out a pilot study, then we choose a simpler (or ‘lower’) scale of measurement and note that the ‘higher’ the scale of measurement, the greater the information content. Finally, measurements of azimuth round the compass can also be made on ordinal or nominal scales, the latter by dividing the compass into sectors.

1.3 Graphical presentation of data

Tabular and graphical presentations of data have two important applications: as a method of familiarisation and experimentation during early stages of a project, or as a method of summary and communication at the reporting stage. Experiments with such presentations at the early stages may reveal previously unsuspected relationships, but we should be aware that tables and graphs can be misleading, intentionally or otherwise, and we should not necessarily be easily convinced that something which appears obvious on a graph must be true. Always be ready to apply a supporting statistical test, even when a graph is highly suggestive and, conversely, do not be deterred from exploring statistical tests if the graph is seemingly disordered. Suitably constructed graphs and tables can be used as a basis for further investigation because many simple statistical tests can be based on them directly. Thus, with experience, casting data into tables and graphs becomes increasingly efficient and purposeful.

1.3.1 Classification

Before tabulation and graphical presentation, data may need to be *formally* classified. Let our elementary understanding of 'classification' be a procedure whereby specimens are allocated to different **classes**, the **class boundaries** between which are defined by us as points on the scale on which the variable is measured. If the variable happens to be measured on a nominal scale, e.g. rock type, then a classification already exists in the form of the 'rules' of rock identification and naming. However, in the case of ordinal and higher scales of measurement, the experimenters must settle on their own definitions of class boundaries. If colour density is important in a study of the weathering of rocks, and has been measured on an ordinal scale, then for the purposes of tabulation or graphing, geologists may find it useful to classify the rocks as 'light', 'medium' or 'dark', erecting class boundaries of their own choosing, but making certain that the boundaries are transferable from one sample (remember the definitions?) to another. In practice, this could be achieved only by laying the several samples side by side. Where measurements are on a ratio scale, class boundaries can readily be placed at convenient values of the variable, but should be defined in such a way as to avoid the difficulty of specimens appearing to fall exactly on boundaries. Thus, a fossil is 50.0 mm long to the limits of precision of measurement. Rather than defining the adjacent classes of '40–50 mm' and '50–60 mm' and thus causing a classification problem, the classes should be defined as 'equal to or greater than 40 mm and less than 50 mm', 'equal to or greater than 50 mm and less than 60 mm'. Now the fossil belongs, unambiguously, to the latter class.

1.3.2 The histogram

This is a commonly used device, but worthy of some scrutiny because of the use that will be made of it. Figures 1.1a–c show three histograms, one each for a nominal, an ordinal and a ratio-scale measurement.

The horizontal scale is divided according to the classification of

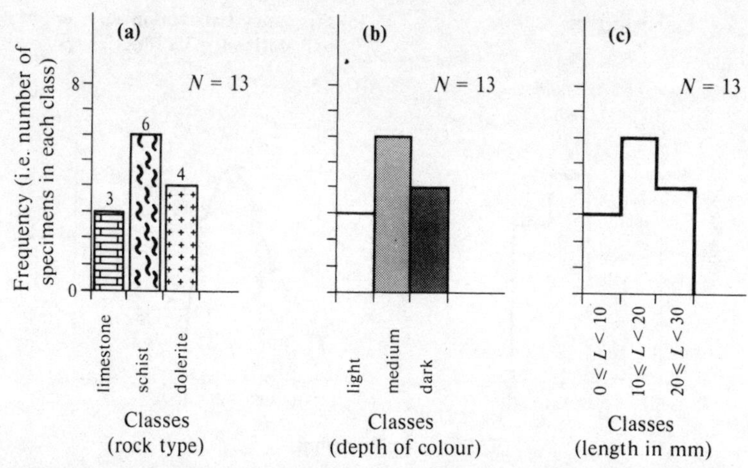

Figure 1.1 Histograms corresponding to (a) nominal, (b) ordinal and (c) ratio scale measurements respectively.

the variable and a box is erected on each class interval. The *area* of the box is made proportional to the number of specimens in that class – the **class frequency**. Note carefully that it is the *area*, not the height, of the box since, in the case of ratio scales, we may choose classes which are not all of the same width. Another point worth noting is that the vertical scale is graduated in the actual frequency, not, as is common, in percentages of the total sample. For much that follows in this book, *percentages may cause problems* and are perhaps best avoided. They are used frequently and loosely and are often allowed to obscure the **sample size** (the number of specimens in the sample) so that virtually all of the appropriate statistical tests in this book are rendered useless. To this end, the histograms above are constructed so that the frequency in each class is readily determined and the sample size (N) is stated. A final subtlety is that for the discontinuous scale (rock type), the boxes are separated, whereas for the continuous scales, the boxes are joined.

1.3.3 The probability density function

If the variable is measured on a continuous ratio scale, then we can derive from the histogram a second type of graph shown in Figures 1.2a and b.

To construct the frequency distribution curve, we mark the centre of the top of each box of the histogram and then draw a smooth curve through these points. If the curve has been drawn carefully,

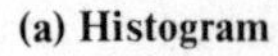

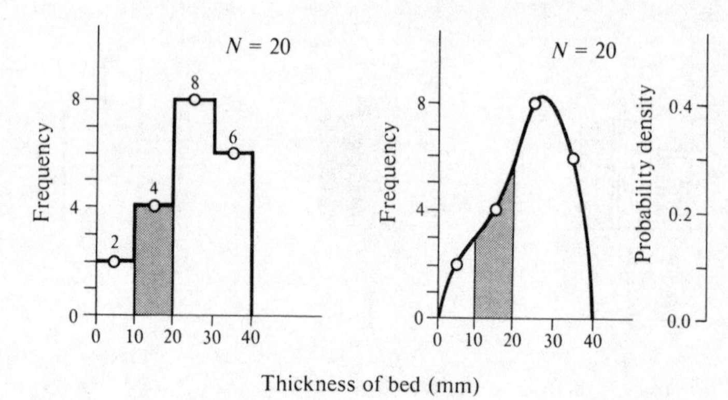

Figure 1.2 The relation between (a) the histogram and (b) the frequency distribution curve or probability density function (pdf).

then the area beneath, and enclosed by the ordinates (verticals) at, say 10 and 20 mm thickness of bed, will be the same as the area of the corresponding box of the histogram (shaded in the diagrams). This property is useful when we remember that the areas of the histogram boxes are proportional to the frequencies of specimens in each class. The total area of the histogram boxes is proportional to the sample size, $N = 20$, made up of 2 in the first class, 4 in the second, 8 in the third and 6 in the fourth, as shown in Figure 1.2a. There are four beds with separate thicknesses between 10 and 20 mm, a proportion of $4/N = 4/20 = 0.20$ of the total sample. Measurement of the corresponding areas beneath the carefully drawn frequency distribution curve should lead to the same result. One advantage of the curve is that we can carry out a similar exercise for *any* range of the variable, totally independent of our original classification, i.e. we can carry out a complete analysis of the frequency distribution in the sample. In brief, the frequency distribution curve can be thought of as a smoothed histogram containing an infinite number of classes, each of infinitesimal width.

The meaning of the alternative name, **probability density function** (pdf), may be illustrated in this way. If we suppose that our sample truly represents the population from which it was drawn, then there are no beds less than 0 mm thick and no beds greater than 40 mm thick. Thus when we choose any single bed from this sample, we can be certain that its thickness will be somewhere in the range from 0 to 40 mm. Such an event, the outcome of which is certain, is said to have a probability (of occurrence) of unity (or 1). On the other hand, an event that never occurs is said to have a probability of zero (or 0). In our example, we shall never find a bed less than 0 mm thick or greater than 40 mm thick. Values of probability intermediate between 0 and 1 are likely to be more interesting. For example, we might have to answer the question 'What is the probability of finding a bed whose thickness falls in the range 10–20 mm?' Of course, we can calculate this from the histogram because the interval 10–20 mm corresponds to one of our classes. The answer is:

(area of shaded box in Fig. 1.2a)/(total area of histogram)
$= 4/20 = 0.20.$

But this coincidence of class boundaries with specified interval is likely to be rare. However, we can use the properties of the pdf for the same purpose. The answer to the same question is:

(area of shaded part of Fig. 1.2b)/(total area beneath pdf)

but the calculation is more cumbersome because we have to find a way of measuring the areas of these awkward shapes. In practice, we should use the **cumulative distribution function** (described in the next section) for this exercise. Note that the greatest probability density (at the summit of the curve in Fig. 1.2b) suggests that the most common bed thickness is about 27 mm.

1.3.4 The cumulative distribution function

There is a third kind of graphical presentation which is extremely useful. Its derivation may be illustrated by Figures 1.3a and b.

First, we imagine constructed a 'cumulative histogram', the area of each box being proportional to the number of specimens in that class *plus* the number of specimens in all other classes to the left, corresponding to *smaller* values of the variable. Using the same data as for the previous, conventional histogram and frequency distribution curve, we obtain Figure 1.3a where for each successive step we add the area of the corresponding conventional histogram box. Next, we relabel the vertical scale, dividing the frequencies by the sample size, to obtain proportions. Finally, putting an open circle on the top, right-hand corner of each histogram box, we draw a smooth curve that is asymptotic to the abscissae (horizontals) at

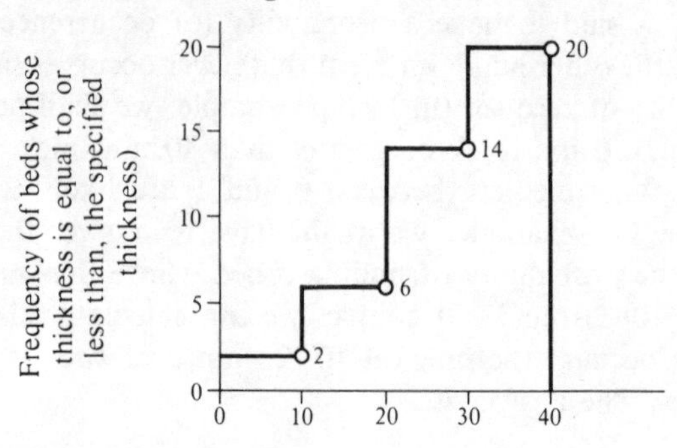

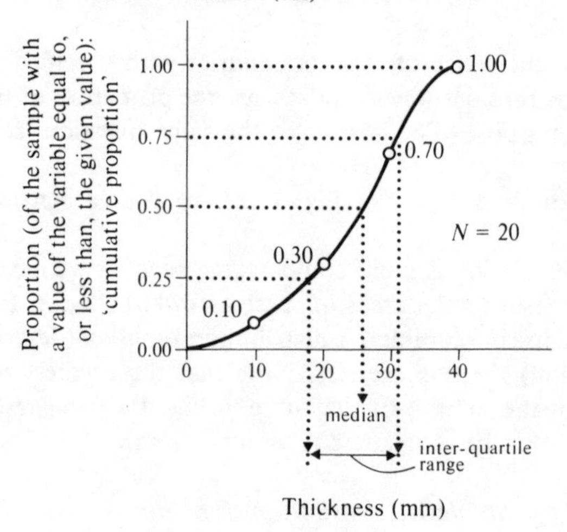

Figure 1.3 The cumulative histogram (a) used to derive the cumulative distribution function (cdf) (b).

values of the proportion of 0 and 1. We have lost all knowledge of sample size because the vertical scale is now expressed in proportions, so this must be stated clearly in the figure: $N = 20$. Note that the height of any given box on the cumulative histogram is equal to the total area of boxes on the conventional histogram that coincide or lie to the left. Similarly, on the cumulative distribution function (cdf), the *height* of the curve corresponding to some specified value of the variable is equal to the *area* beneath the pdf to the left of the corresponding ordinate. We can use this feature to answer more easily the question 'What is the probability of find-

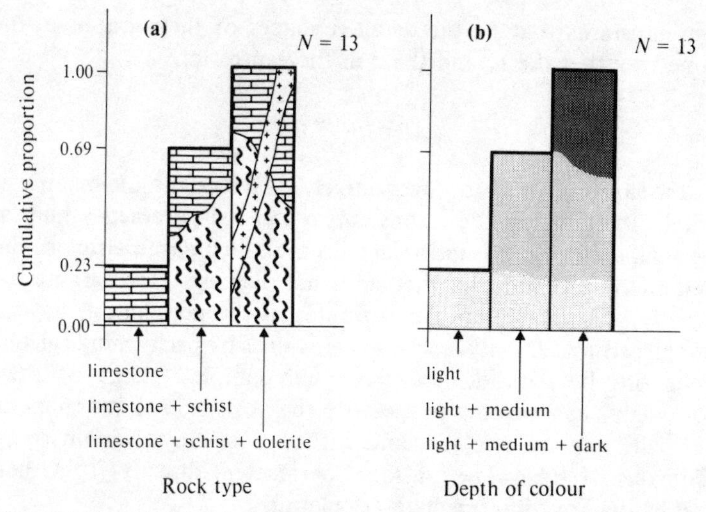

Figure 1.4 Cumulative distribution functions constructed for (a) nominal and (b) ordinal scale variables. Because the variables are discontinuous, the functions are step-like.

ing a bed between 10 and 20 mm thick?' Using Figure 1.3b, we note that beds of thickness 20 mm or less, constitute 0.3 of the total sample, whereas beds of thickness 10 mm or less, constitute 0.1 of the total sample. The required probability is thus:

$$0.3 - 0.1 = 0.2.$$

Lastly, it should be noted that discontinuous, as well as continuous, variables can be graphed in this way. Thus, bizarre as they may seem, the two cumulative distribution 'step' functions of Figures 1.4a and b have sound practical application, which will be discussed in Chapter 3.

1.3.5 Parameters of distributions of linear data

We noted in the Introduction that numerical parameters can be calculated from our data, mostly via simple mathematical expressions, and then used to give concise descriptions of features of our data distributions. We shall see elsewhere (especially in Chs 5, 8 & 9) that these parameters arise because it is often convenient to suppose that our *sample* data is drawn from a *population* in which the distribution of the data follows some mathematical expression, called a **distribution function**. These mathematical expressions contain constants that fix the detailed shapes of their graphs in the same way that the m and the c in the expression:

$$y = mx + c$$

fix the slope and position, respectively, of a straight line on an x–y graph. In statistics, these constants are called **parameters** and (as we shall see in Ch. 5) statistical tests are called **parametric** or **non-parametric** according to whether or not they use these parameters. Strictly, the parameters are attributes of the population, not the sample, so we can only *estimate* their values by performing calculations on the data derived from our samples. Some of these parameters can be used to quantify the position and the shape of both the pdf and the cdf, and so it is possible to obtain rough estimates of the values of some parameters directly from these graphs, without intermediate calculation.

The *position* of the pdf or the cdf, or more particularly, the *central value* of the variable in whose distribution we are interested, can be specified by any one of the three following parameters.

(a) **Mode:** the value of the variable corresponding to the summit of the pdf and nearly always estimated graphically. (Approximately 27 mm for the data of Figure 1.2b.) For classified data presented on a histogram, the **modal class** is that in which the frequency is greatest: schist or medium in the cases of Figures 1.1a and b respectively. The data that arise in some samples, even of large size, may be polymodal, i.e. exhibit more than one summit on the pdf or more than one class on the histogram, with frequency greater than that of its immediate neighbours. However, in Section 1.3.8 we shall see, in the discussion of circular histograms, that such local modes may simply be a function of our arbitrary choice of class boundaries.

(b) **Median:** the value of the variable such that half the sample values are greater and half smaller; on the pdf this is the ordinate (vertical) that cuts the area beneath the curve into two equal parts. Calculation according to the first of these two definitions is straightforward if the specimen values of the variable have been **ranked**, i.e. placed in order of increasing magnitude. If the number of specimens is odd, then the median value of the variable is that associated with the specimen that is half way along the ordered list. If the number of specimens is even, then the median value is taken as the average of the two specimen values that flank the halfway mark. Direct calculation according to the second definition is difficult, but the required median value can be obtained directly from the cdf by finding that value of the variable that cuts off a proportion of 0.50 of the sample. In Figure 1.3b, the median value is approximately 26 mm.

(c) **Mean:** the usual sense associated with average, i.e. the value of the variable obtained by adding together the values of the variable measured on each of the specimens and dividing by the total number of specimens. In formal mathematical terms:

$$\overline{M} = (X_1 + X_2 + X_3 + \ldots + X_N)/N \tag{1.1}$$

where $\overline{M}$ is the mean, X_1, X_2, X_3 ... X_N are the variable values attached to specimens numbered 1, 2, 3, ... N respectively, and N is the sample size, i.e. the number of specimens in the sample. As an aside, let me take advantage of this simple example to show how Equation 1.1 may be shortened by using the summation symbol Σ:

$$\overline{M} = \sum_{i=1}^{i=N} (X_i)/N \tag{1.2}$$

in which $\sum_{i=1}^{i=N} (X_i)$ represents $(X_1 + X_2 + X_3 + \ldots + X_N)$.

The subscript i is used as a general label for specimens and the range of values it is to take is specified below and above the Σ by $i = 1$, the lowest, in steps of 1 up to $i = N$, the highest. The Σ is a summation symbol which invokes the **summation convention**, instructing us to add together expressions of the type that appear in the brackets that immediately follow, for all possible values of the subscripted index i. In this application, the expression (X_i) is simple: just the value of a variable, but later we shall use the summation symbol with more complicated expressions. Lastly, note in Equation 1.2 that the summation must be completed before dividing by N – the N is *outside* the brackets.

I have already noted limitations involved in the estimation of some of these parameters above. Another springs from the fact that, in many cases, we are trying to estimate a population parameter from measurements made on a sample of restricted size. As a result, we need to give some thought not only to the ease of calculation, but also to the **stability** of the estimate. For example, a value for the mean calculated by Equation 1.1 may be seriously disturbed if the sample contains one or two specimens with extreme values of the variable, whereas the median is much less influenced. Ease of calculation may depend on circumstances. Counting halfway along a line of loose pebbles, arranged in order of increasing length, immediately locates the pebble possessing the median length, but if the same pebbles are firmly cemented into a conglomerate, then calculation of the median is tedious compared with that of the mean.

The *shape* of the pdf or cdf can be described in terms of two groups of parameters, related to **dispersion** and **skewness**. Quantifying the dispersion allows us to specify how widely the specimen values are spread on either side of their central value (i.e. mean or median or mode). A large value for dispersion is associated with a wide, rather flat shape for the pdf, with a mode that is difficult to locate with confidence, whereas a small dispersion value corresponds to a high, sharply peaked pdf with a mode that is easily located. A simple and sometimes effective dispersion parameter is the **range**, calculated by subtracting the smallest specimen value from the largest. Unfortunately, it is easy to see disadvantages arising from chance collection of extreme specimens: the resulting value for the range will be severely affected.

Rather less susceptible to idiosyncrasies of sampling is the **interquartile range** – easily estimated from the cdf. Figure 1.3b illustrates the procedure: ordinates (verticals) are found that cut off 0.25 (lower quartile) and 0.75 (upper quartile) of the distribution respectively, i.e. one quarter and three quarters of the total sample. The distance between these ordinates, as measured in the units of the horizontal scale, is the interquartile range. In this case: $31 - 18 = 13$ mm. Sometimes the value is halved to define the semi-interquartile range. Either parameter is clearly much more stable under sampling fluctuations than the (overall) range.

Neither range nor interquartile range is particularly amenable to statistical testing. For these purposes, by far the most useful measure of dispersion is the *standard deviation*, the estimation of which is dealt with in Section 5.2.2 because it is also a parameter of the important theoretical distribution developed by Gauss. Although useful in testing, we have to note the labour involved in measurement (every specimen on a ratio scale) and the labour of calculation required for its estimation.

Another aspect of the shape of a naturally observed distribution concerns the symmetry (or asymmetry) of its pdf. In many cases, we anticipate that the pdf will be single-peaked (i.e. unimodal) with slopes and tails approximately symmetrically arranged about the mode. Any tendency for the pdf to lean to one side introduces the idea of asymmetry, with the mode displaced towards either smaller or larger values of the variable. This characteristic is called skewness, and two measures of the property are:

$$\text{skewness} = \frac{\text{upper quartile} + \text{lower quartile} - 2 \times \text{median}}{\text{interquartile range}} \tag{1.3}$$

$$\text{skewness} = \frac{3 \times (\text{mean} - \text{median})}{\text{standard deviation}} \tag{1.4}$$

Both expressions yield zero for symmetrical pdf's and have a total range from -1 (if the pdf leans towards higher values) to $+1$ (if the opposite holds). However, for specific samples, the expressions give slightly different estimates, so always state which equation has been used.

Moderately to highly skewed samples can cause problems when subjected to more refined statistical analysis where an assumption of symmetry of the pdf has to be made. A common solution to such difficulties is to **transform** the original measurements by basing the analysis not on the original values of the variable, but on some simple function such as the logarithm or square root, etc. The objective then is to find a suitable transformation that reduces the skewness of the pdf. Sometimes the context of the sample suggests a suitable transformation; otherwise a variety may be tried empirically.

1.3.6 Orientation data

Orientation data arise commonly in geology in two-dimensional and three-dimensional situations, i.e. on the circle and on the sphere, and thus we may talk of **circular** and **spherical** distributions, in addition to the linear distributions discussed above. Graphical presentation of spherical distributions is nearly always via the stereographic or some similar projection, in which orientations appear as points. There are various ways of contouring the density of these points, but all except one of them (referred to in Ch. 9) lead to no further practical advantage with respect to statistical testing: they are laborious and serve little purpose beyond enhancing the graphical presentation.

1.3.7 The circular plot

Corresponding to the stereographic projection in three dimensions, we have the circular plot in two dimensions. In the example shown in Figure 1.5, points on the circumference of the circle represent the directions of gape of the apertures of 36 goniatites on a bed. Such a plot is rapidly prepared and gives a good visual impression of density variations. It can also be used directly to find a **circular median direction** by searching for a diameter that divides the sample into two equal parts (in this case, 18 points on either side of the required diameter) and then choosing the direction along that diameter in which point density is greater.

1.3.8 The circular histogram

Circular histograms may also be constructed (Figs 1.6a & b). These two histograms both display the goniatite data above, and both

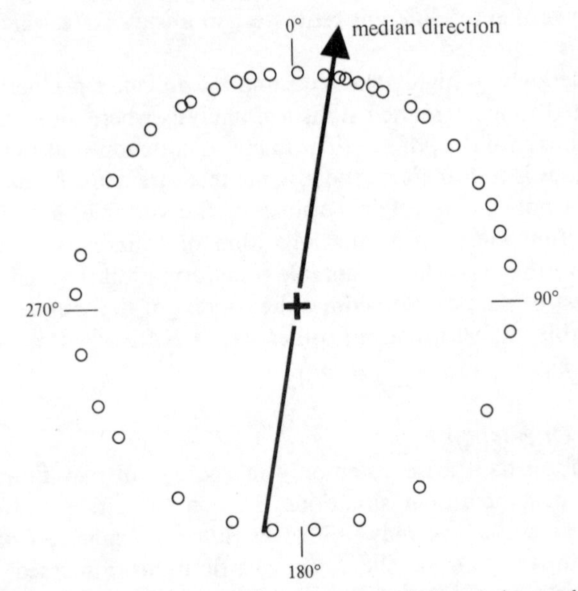

Figure 1.5 A circular plot of the directions of gape of 36 goniatites on the surface of a bed in the North of England (data kindly provided by Dr W. B. Heptonstall).

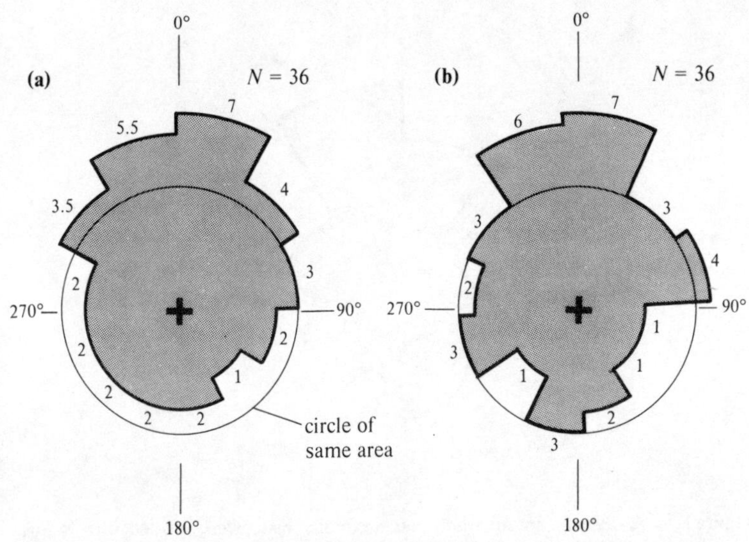

Figure 1.6 Circular histograms for the same data as in Figure 1.5. These two diagrams differ principally in the choice of class boundaries.

have a class width of 30°. The area of each sector of the diagram is made proportional to the class frequency (which is marked at the perimeter) so that the radius of the sector is given by:

$$r^2 = (2Af)/(Nw) \tag{1.5}$$

where r is the sector radius, A is the total area of the histogram, f is the class frequency, N is the sample size and w is the class width in radians (1 radian = $180/\pi$ degrees).

Thus we see that some computational effort is needed to construct the circular histogram, and its step-like discontinuities detract from the continuous variation that is brought out by the circular plot. However, the modal class (i.e. the class with greatest frequency) is easily found. Nevertheless, one serious disadvantage, as with all histograms, is that the appearance can be radically altered by the choice of the *positions* of the class boundaries. Although both diagrams above have a class width of 30°, that on the left has the first boundary set at the origin (0°), whereas that on the right has boundaries set such that the modal class is centred on the median direction. This relatively minor change, involving only a 10° anticlockwise shift of the class boundaries, has had a marked effect on the overall appearance of the diagram. Histograms, whether

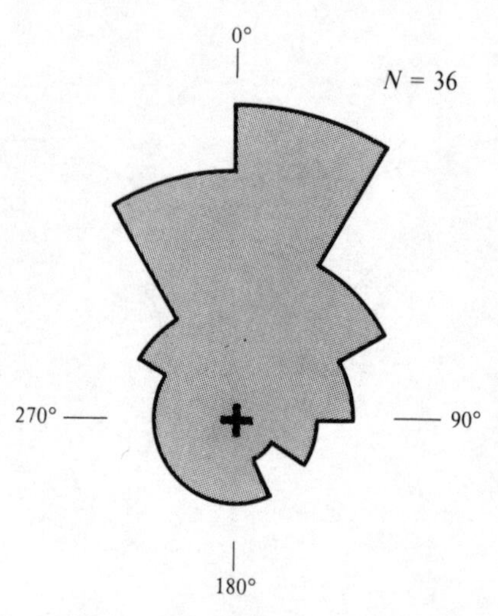

Figure 1.7 A diagram masquerading as a circular histogram, plotted with radius (not area) of each sector proportional to the class frequency. Same data as Figure 1.5.

linear or circular, are prone to this kind of instability, especially when sample sizes are small, and so should not be discussed too casually or given too much weight.

Another common difficulty with the circular histogram is that it is often thoughtlessly constructed with the *radius* of the sector proportional to the class frequency, rather than the *area*, as above. This leads to gross distortions, as shown by Figure 1.7. The eye is now likely to be unduly impressed by the dominating size of the modal sector, completely out of proportion to its true significance. The degree of preferred orientation is greatly overemphasised by this kind of plot.

1.3.9 Other circular diagrams

Other methods of representing circular data may be devised, but they rarely approach the circular plot in speed of preparation and usefulness. Thus, the circular histogram may be 'unrolled' and presented in a pseudo-linear form with 0° at one end of the horizontal axis and 360° at the other end, but this masks the essential cyclicity of the data. Also, a circular analogue of the frequency-distribution curve may be constructed, but little practical advantage arises.

Lastly, we should note that circular data may be axial as well as directional (remember the difference?). Fossil plant stems of

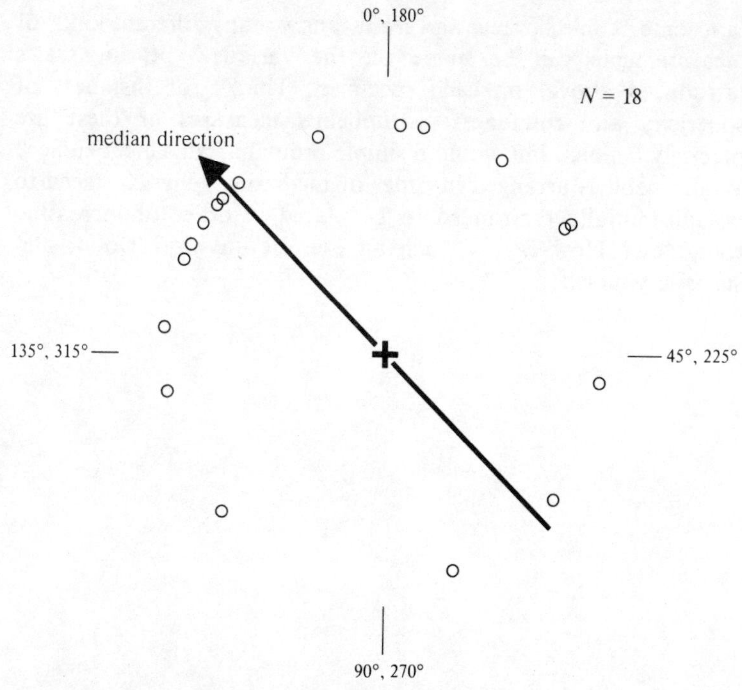

Figure 1.8 A circular plot of fossil plant stems. These specimens are axial: one end is indistinguishable from the other, so only 180° of the compass are needed to specify their orientation. However, to emphasise the cyclicity of the data and render greater practical usefulness to the diagram, the orientation angles have been doubled (data kindly provided by Dr W. B. Heptonstall).

uniform width may have one end indistinguishable from the other, and therefore we need use only 180° of the compass to specify their orientations. Again, bizarre as it may seem, there are several practical advantages in doubling these angles and presenting them on a circular plot as in Figure 1.8. Such a diagram may have north and south at the top, and west and east at the bottom(!) but one circuit represents one cycle of the data and this is the important advantage: the plot can be used directly in statistical testing, as in Chapter 8.

Exercise. Pebbles are conveniently accessible geological objects which have already featured in a number of illustrations and will do so again. However, please do not view this book as an exercise in pebble-counting; you will find that all sorts of geological problems will yield to the methods outlined here. By way of exercise, obtain some pebble samples and see just how many different kinds of measurements can be made on the various types of scales introduced above, on each specimen. Think, for instance, of sphericity and roundness: Krumbein's measures of these are precisely defined, but would a simple ordinal arrangement suffice? Would pebbles arranged in order of increasing sphericity need to be substantially rearranged to be placed in order of increasing roundness? How easy is such an exercise anyway? Do it, and surprise yourself!

2 Coin tossing and stratigraphy

This chapter sets out some elementary ideas of probability and hypothesis testing using, by way of illustration, the familiar laws of chance associated with coin tossing. It is an important chapter because the theme of hypothesis testing runs throughout the book. Although the contents of this chapter have some applications in poker and other games in which chance is an important element, I shall end by showing how the whole structure may be moved without change into the realm of stratigraphy.

2.1 A coin-tossing experiment

Suppose we have a sample of identical coins, each of which has one side named 'head' (h) and the other side named 'tail' (t). When we toss such a coin, we expect the probability of its landing h uppermost to be the same as the probability of its landing t uppermost. In other words, if H is the probability of heads and T is the probability of tails, then we expect $H = T = \frac{1}{2}$. For the time being however, let us lay aside the assumption that the coins are unbiassed: because of some manufacturing defect, they are all biassed such that H is not equal to T. Nevertheless, by our understanding of probability, $H + T = 1$, i.e. a coin *must* come to rest showing either h or t (we thus discount the perhaps devious suggestion that a coin comes to rest on its edge).

Now consider a trivial experiment in which we toss a single coin. There are two possible outcomes: t with probability T, or h with probability H. The results of the experiment can be set out as in Table 2.1, in which each column corresponds to a possible outcome of the experiment.

Table 2.1 Results obtained by tossing a single coin.

possible outcome number	1	2	$(=2^1)$
coin 1 lands	t	h	uppermost
probability of outcome	T	H	
conclusion	the probabilities of 0 or 1 heads are T (outcome 1) and H (outcome 2) respectively		

Table 2.2 Results obtained by tossing two coins.

possible outcome number	1	2	3	4	$(=2^2)$
coin 1 lands	t	t	h	h	uppermost
coin 2 lands	t	h	t	h	uppermost
probabilities	T^2	$T\times H$	$H\times T$	H^2	
conclusion	the probabilities of getting 0 or 1 or 2 heads on tossing two coins are T^2 (outcome 1), $T\times H+H\times T=2\times T\times H$ (outcomes 2 and 3) and H^2(outcome 4) respectively				

We now refine the experiment by adding a second coin and the results are displayed in Table 2.2. There are two points we should note to ensure that we understand this table. First, consider the probabilities of outcome. There are two coins, they are tossed independently, and there is no way that the landing of one coin can affect the landing of the other. Consequently, the probability of coin 1 landing and showing *t* is independent of the probability of coin 2 landing and showing *t*, or any other way. In short, the two events are independent, and the rule for combining probabilities of

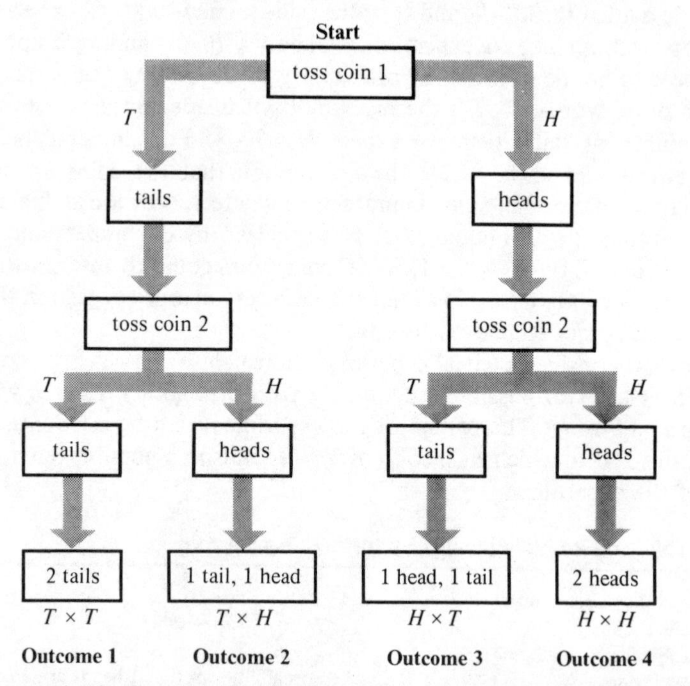

Figure 2.1 All possible sequences of events in a tossing experiment with two coins. T and H attached to arrows represent the probabilities of 'tails' and 'heads' respectively. Overall probabilities are given at the end of each of the four branches.

independent events is to multiply them (e.g. Sprent 1977, p.32). This should be clear from the alternative presentation in Figure 2.1. Starting at the top, tossing coin 1 immediately divides further events into two sets, a fraction T for which coin 1 landed 'tails' and a fraction H for which it landed 'heads'. Of this fraction T, further subdivision takes place on tossing coin 2 with fractions T and H of the original fraction T. The rule for calculating a fraction of a fraction is again, to multiply, thus deriving the overall probabilities of outcomes as displayed at the bottom of Figure 2.1.

Secondly, as we are interested only in the *total* number of heads that results in any particular outcome, we can calculate the probabilities of getting 0, 1, or 2 heads by combining the probabilities of the separate events, as above. These are T^2, $2 \times T \times H$, and H^2 respectively, and we note that we can generate these overall probabilities by multiplying out ('expanding') the expression $(T+H)^2$ to give

$$(T+H)^2 = T^2 + 2 \times T \times H + H^2$$

This is the **binomial** expansion, and probabilities thus calculated are called **binomial probabilities**. They govern all processes at any stage of which only one of two possible events may occur.

We are ready now for a further elaboration of the experiment, three coins: the results appear in Table 2.3.

The way ahead is now clear, I hope. If we have an experiment with N coins, then the probabilities of 0, 1, 2, . . . N heads are successive terms in the expansion of $(T+H)^N$. This is the *binomial* expansion and its *general term* is given by:

$$\begin{aligned} \text{probability of } R \text{ heads} &= \binom{N}{R} \times H^R \times T^{(N-R)} \\ &= \binom{N}{R} \times H^R \times (1-H)^{(N-R)} \end{aligned} \qquad (2.1)$$

Table 2.3 Results obtained by tossing three coins.

outcome	1	2	3	4	5	6	7	8	$(=2^3)$
coin 1	t	t	t	t	h	h	h	h	
coin 2	t	t	h	h	t	t	h	h	
coin 3	t	h	t	h	t	h	t	h	
probabilities	T^3	$T^2 \times H$	$T^2 \times H$	$T \times H^2$	$T^2 \times H$	$T \times H^2$	$T \times H^2$	H^3	
conclusion	the probabilities of 0, 1, 2, 3 heads on tossing 3 coins are T^3, $3 \times T^2 \times H$, $3 \times T \times H^2$ and H^3 respectively, terms in the expansion of $(T+H)^3$								

$\binom{N}{R}$ is the number of combinations from N objects taken R at a time and is given by:

$$\binom{N}{R} = \frac{N!}{R! \times (N-R)!}$$

where ! is the 'factorial' sign ($N! = N \times (N-1) \times (N-2) \ldots 3 \times 2 \times 1$; e.g. $4! = 4 \times 3 \times 2 \times 1 = 24$).

Example

What is the probability of getting 4 heads in tossing 6 fair coins (taking 'fair' to mean $H = T = \frac{1}{2}$)? In this example, $N = 6$ and $R = 4$, so:

$$\binom{N}{R} = \frac{6!}{4! \times (6-4)!} = \frac{6 \times 5 \times 4 \times 3 \times 2 \times 1}{4 \times 3 \times 2 \times 1 \times (2 \times 1)} = \frac{6 \times 5}{2 \times 1} = 15$$

Also:

$$H^R = (\tfrac{1}{2})^4 = \tfrac{1}{2} \times \tfrac{1}{2} \times \tfrac{1}{2} \times \tfrac{1}{2} = 1/16$$

and

$$T^{(N-R)} = (\tfrac{1}{2})^2 = \tfrac{1}{2} \times \tfrac{1}{2} = 1/4$$

So the required probability is $15 \times (1/16) \times (1/4) = 15/64 =$ approximately 0.234.

Suppose we have an experiment with 14 coins and we *assume initially* that the coins are fair (i.e. $H = T = \frac{1}{2}$). Using Equation 2.1, we can tabulate the expected probabilities of getting 0, 1, 2, ... 13, 14 heads in the experiment when all the coins are tossed together. This table appears in Figure 2.2 with a graph of the same numbers (a pdf analogous to Fig. 1.2b, but with variable R discontinuous in that it takes only integer values, so that the resulting function is stepped).

If our coins are fair, then we expect to get heads and tails in approximately equal abundance. On the other hand, if we get very few heads or very many heads, we might conclude that:

either (a) our assumption $H = T = \frac{1}{2}$ is correct but a *rare event* has occurred.

or (b) our assumption $H = T = \frac{1}{2}$ is incorrect.

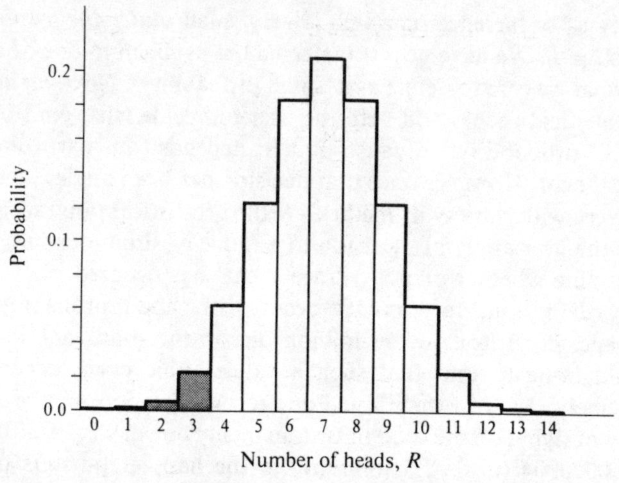

Figure 2.2 The probability density function for a binomial distribution in a sample of size $N = 14$ with $H = T = \frac{1}{2}$.

Number of heads	*Probability*	*Number of heads*	*Probability*	*Number of heads*	*Probability*
0	0.0000610	5	0.122	10	0.0611
1	0.000854	6	0.183	11	0.0222
2	0.00555	7	0.209	12	0.00555
3	0.0222	8	0.183	13	0.000854
4	0.0611	9	0.122	14	0.0000610

Which of these two we choose will be dictated by our understanding of what constitutes a 'rare event' (discussed immediately below). Meantime, as a corollary to (b), if we have very few heads then we might be persuaded to assume that H is less than T and conversely, if we have very many heads, then we might assume that H is greater than T.

2.2 'Small probabilities' and 'rare events'

The problem now is: how do we quantify what we mean by *very few* heads or *very many* heads? Expressed differently: at what *critical value* of R (the number of heads) do we decide to reject the hypothesis that $H = T$, and accept the alternative that H is not equal to T? The answer must be: at some value of R, the probability of occurrence of which is very small *under the assumption* that $H = T$. We have now transferred the problem to one of *choosing* what we care to define as a 'small probability'. Now, *no amount* of statistical theory will help the user to decide what constitutes a 'small probability' because much depends on particular circumstances. However once that decision has been made, then there is a very wide variety of methods (called statistical tests) which will help the user apply his decision to a range of situations in order to determine whether or not a 'rare event' has occured.

By definition, **rare events** occur with small probability. On average, 20 Britons of 50 million die on the road each day. We would hope to find that such a catastrophic event occurs with extremely small probability. For the average person, the probability of dying on the road in Britain in any one day is 20/50 000 000 = 0.0000004 (or 1 : 2.5 million). In the happier pursuits of coin tossing and other gambling pastimes, we are likely to take a risk when the chances of winning are, say, 1 : 20 or 1 : 100, but possibly not much less than this. In short, our notion of a 'small probability' is going to depend on the particular situation. In the case of the coin-tossing experiment with 14 coins, suppose we decide that a 'small probability' is 1 : 20 (0.05) or less. If, on tossing all the coins once only, no heads result ($R = 0$), then a rare event has occurred (assuming 'fair' coins) because the table in Figure 2.2 tells us that the associated probability is 0.0000610 (i.e. approximately 1:16400). If, on repeating the experiment, we consistently get 0 or 1 heads, then again a rare event is occurring (still assuming 'fair' coins) because the probability of getting 0 *or* 1 heads is the sum of the separate probabilities (0.0000610 + 0.000854 = 0.000915, approximately 1 : 1100). To calculate the 'critical value' of the number of heads, equal to or less than which a rare event is still occurring, we simply continue to add together the successive probabilities in the table in Figure 2.2 to derive Table 2.4. From this, we see that a result of 3 heads or fewer complies with our chosen definition of a rare event.

Table 2.4 Calculation of the critical value of R.

probability of	0 heads	0.0000610
	1 or fewer heads	0.000915
	2 or fewer heads	0.00646
	3 or fewer heads	0.0287
		0.05 (our 'small' probability)
	4 or fewer heads	0.0898

This procedure is very important so let us begin to formalise what we are doing:

(1) We assume that the coins are fair so that $H = T = \frac{1}{2}$. If the coins are not fair, then they are biassed on tossing to produce a smaller number of heads than tails.
(2) We choose to test the assumption by carrying out the experiment of tossing a sample of coins and counting the number of heads that results.
(3) We choose to toss 14 coins and we define a 'rare event' to be one that occurs, at most, 1 in 20 times (i.e. with probability 0.05 or less).
(4) Since we are going to base our conclusion on the number of heads R that result, we need to know the probabilities of obtaining specific values of R. The probabilities have been calculated above.
(5) Table 2.4 tells us that if 3 heads or fewer result, then a 'rare event' has occurred under our assumption (1) above.
(6) Finally, we conduct the experiment and conclude as follows:
    (a) If R is 4 or more: accept that the coins are fair and that H is equal to T.
    (b) If R is 3 or less:
        either (i) accept that the coins are fair and excuse the result with the argument that a 'rare event' has indeed taken place;
        or (ii) reject that the coins are fair and accept that the coins are biassed to produce a smaller number of heads than tails when tossed.

2.3 The six steps of hypothesis testing

Siegel (1956, pp. 6–17) formalised these six steps of hypothesis testing as follows, but some of the ideas I introduce here are in outline only because I want to return to them for fuller discussion in Chapter 10, once I have described what certain statistical tests can do.

2.3.1 State the 'null' and 'alternative' hypotheses

The **null hypothesis** (usually symbolised H_0) is a hypothesis of no differences, sometimes set up with the express purpose of being rejected in favour of the **alternative hypothesis** (symbolised H_1) which may be a statement of the observer's 'working hypothesis'.

In the example above, H_0 is that the coins are fair, there is no difference between the probability of 'heads' and the probability of 'tails', and so the null hypothesis embraces the suggestion that $H = T = \frac{1}{2}$. The alternative hypothesis is that the coins are not fair; they are all imbalanced (or biassed) in some way so that when tossed, they tend to fall with tails uppermost. The probabilities are then such that $H < T$.

2.3.2 Choose the statistical test

A large number of statistical tests may be available in a particular situation. There is a measurement requirement and a statistical model associated with each test which together imply some assertions about the sampled population, the manner of sampling and the method of measurement. Sometimes we can test whether or not the conditions of a particular model are met but, more often, we have to make some assumptions. Consequently, the statistical model is defined in part or in whole by a list of assumptions and a statement about the supposed validity of the measurements, whatever the situation. However, we note the general rule that the most 'powerful statistical tests have the most extensive assumptions (I postpone discussion of the 'power' of statistical tests until Section 2.4). Basically, we search for overall efficiency and, in particular, full use of the data, and study carefully the limitations of the test.

In our example, we have constructed our own statistical test based on a study of the binomial distribution. More often, however, we are likely to make use of a 'standard' test and naturally, as our experience and knowledge of these gradually expand, so we find it easier to make the choice (more will be said about this in Ch. 10, after we have gained a little of this experience).

2.3.3 Choose the size of the sample N and define a 'small probability' a

Again, the choice of sample size N is likely to be influenced by our choice of statistical test (or vice versa), and the context of the application is certainly going to influence both these choices. We are likely to feel more naturally confident in our conclusions the larger our sample size becomes, and it is certainly true that the probability of making mistakes falls, but not necessarily dramatically, with increasing sample size. However, one of the important points in formalising a hypothesis-testing procedure lies in trying to achieve a worthwhile objective with the minimum of effort or at minimum cost. Once we have seen some statistical tests in action, we shall note how we can predict, before applying a test, the likely effort (or cost) in relation to our specified objective. In cases where such predictions are difficult, we can always probe the situation by choosing a simply applied, less 'powerful' test that involves less effort. The outcome of this 'ranging shot' is valuable in setting up the main test. In our present example of coin tossing, choosing a sample size of $N = 14$ perhaps opens us to making mistakes, but it so happens that $N = 14$ produces a set of convenient probabilities for your exercise to come!

As discussed above (Section 2.2), only *you* can choose to define what you are going to accept as constituting a 'rare event'. Statistical theory offers only marginal help here because we are entering the field of decision theory where factors of cost, effort, and even personal reputation, enter into the development, but this is largely beyond the scope of this book. One thing we can assess (as we shall see shortly in Section 2.4) is the probability of our making a mistake in drawing our conclusions. This represents a valuable achievement. In the current example, I have chosen a 'small probability' to be $a = 0.05$ so that the chances of the 'rare event' of 14 fair coins giving 'very few' heads when tossed is going to be 1 in 20, or less.

2.3.4 *Evaluate or otherwise discover the frequency distribution of the 'test statistic'*

We define a **test statistic** as the number we observe or calculate, having completed an experiment and with which we enter a statistical test. The documentation or derivation of the statistical test will include reference to the procedure by which the numerical value of the test statistic is found. ('Test statistic' is not to be confused with 'statistical parameter'.) In our current example, the number of heads R is the test statistic, found by counting. With other tests, the test statistic may be found graphically or by application of a mathematical formula.

In order to decide which values of our test statistic are 'common' and which are 'rare', under our null hypothesis *true*, we need to know the frequency distribution (or probability density function) of the test statistic, or at least the 'critical value' at the boundary between 'common' and 'rare'. In the example, the probability of occurrence of all possible values of the test statistic has been calculated and presented graphically (Fig. 2.2). In general, we *may* be able to calculate such probabilities, but we are much more likely to be interested in no more than the 'critical values' of the test statistic and, for most standard tests, these values are tabulated or graphed. Such tables or graphs are nearly always all we need for this stage of the hypothesis-testing procedure.

2.3.5 Define the 'critical region' (or the 'region of rejection')

The **critical region** is the region of the frequency distribution of the test statistic which contains those *extreme* values of the test statistic associated with which, *under* H_0 *true*, there is a probability of occurrence of a or less.

Referring to Table 2.4, Figure 2.2 and drawing an analogy with our discussion of Figure 1.2a, this cryptic definition can be illustrated thus. We want to find the extreme values of the number of heads R associated with which, *under* H_0 *true*, there is a probability of occurrence of 0.05 or less. Table 2.4 in effect tells us how the fractional area under the left-hand tail of the 'curve' of Figure 2.2 increases as we add successive boxes. Note that our critical region will lie under the left-hand tail because our alternative hypothesis H_1 says the coins are biassed to land 'tails' uppermost, thus producing few 'heads'. Adding together the boxes for 0, 1, 2 and 3 heads gives an area of 0.0287 of the total area beneath the 'curve'. Adding the box for 4 heads to this increases the figure to 0.0898 and thus exceeds our chosen value of 0.05. Thus, the critical region is that shaded in Figure 2.2 and may be spoken of as 'containing' values of R of 3 or less. The 'small probability', a, which I have chosen to call the **size of the critical region,** is also known as the 'level of significance', a usage arising from the common problem that we are searching for a 'significant' difference between fact and theory or between two or more sets of observations.

2.3.6 The decision

If the 'experiment' gives a value of the test statistic in the critical region, then H_0 is rejected. There are two possible explanations:

(a) H_0 is actually true and a rare event has occurred.
(b) H_0 is false.

In reporting our decision, we should *never* omit reference to the size of critical region used.

2.4 The probability of being wrong

Note very, very carefully that the statistical procedure of hypothesis testing does not *prove* anything! Indeed '*statistics prove nothing*'. Even if the experiment yields $R = 0$, we have still not *proved* that the coins are biassed, in the legal or mathematical sense of the word. The statistician will always admit that his conclusions may be wrong, but differs in that he will always be able to assess the *probability* of his conclusions being in error. There are only two kinds of error that can be made, as may be illustrated by Table 2.5. Our six-step procedure is designed to detect rare events under the hypothesis H_0 *true*. On the occurrence of a rare event whose probability is a, we reject H_0 and conclude H_1 *true*, even though, *in fact*, H_0 *may* be true. This being the case, we have fallen into a 'type I error', i.e. we have rejected the null hypothesis when, in fact, it is true, and the probability of doing this is clearly a. We may also fall into a 'type II error', namely, to accept the null hypothesis when, in fact, it is false. The probability of doing this is symbolised by b. a and b are not independent: they are linked via the sample size N in a 'see-saw' relation that differs in detail from one statistical test to the next. Thus for a given test and a given sample size N, as we allow a to decrease, so b increases and vice versa. The mathematical nature of the relation is beyond the scope of this book (but see Meyer 1975, pp. 342–9) though we can work to the guidelines in the next paragraph. Generally, we like a and b to be about equal. However, to make them both arbitrarily small (so as to reduce the probabilities of types I and II errors), we need to increase N. Only when N = the whole population does $a = b = 0$.

In most of the tests described in this book, and for most of the sample sizes practicable in many exploratory geological circumstances, we can choose values of a of the order of 0.05 to 0.01; two commonly used values but possessing no magical virtues. We are then reasonably confident that b is of roughly the same size as a. We note that statistical tests are said to be more 'powerful' if they are associated with smaller values of a and b for a given value of N. Given a more intimate knowledge of our statistical tests than developed here, we can specify both a and b in advance and then

Table 2.5 Probabilities of errors in hypothesis testing: I, type I error; II, type II error.

		Fact	
		H_0 true	*H_1 true*
Decision	accept H_0	$1 - a$	b (II)
	accept H_1	a (I)	$1 - b$

calculate the required sample size N. Under such conditions, we then define the power of the test to be $1 - b$, the probability of accepting H_1 when indeed it is true.

2.5 The nature of the alternative hypothesis: one-tail and two-tail tests

One final point remains for discussion: the exact nature of the alternative hypothesis H_1 has an influence on the *location* of the critical region. Above, our alternative hypothesis H_1 has been that the coins are biassed towards producing fewer heads than tails. Suppose it had been simply that the coins were biassed, *without specifying the direction* of the bias. Amongst the rare events under the null hypothesis true, we now have to include the possibility of the experiment producing 'very many' heads as well as 'very few' heads. Clearly, our critical region now has to be shared *equally* between the two 'tails' of the 'curve' in Figure 2.2. Still keeping $a = 0.05$, we need half ($a/2 = 0.025$) under one 'tail' and half under the other. Since the 'curve' is exactly symmetrical, we can use Table 2.4 to show that the critical region now contains values of R of 0, 1, 2, 12, 13, 14 and that the critical values of R are 2 or less and 12 or more. Thus, depending on whether or not the alternative hypothesis has an inference of 'direction' in its wording, so we apply either a **one-tail** test or a **two-tail** test.

Exercise. In the coin-tossing experiment above, find the critical values of R in the one-tail and two-tail situations for a critical region of size 0.01. Find the answer in Table 2.6.

2.6 An application in stratigraphy

To round off this chapter, I want to transfer the whole of the theoretical discussion above into the realm of practical stratigraphy. Suppose a geologist has mapped an area in which a certain part of the stratigraphical succession is exposed in several more or less parallel stream beds. In many of the stream beds, the geologist notes that a certain fossil disappears from the stratigraphical record at a level that differs from one stream to the next, presumably in part due to accidents of preservation. Let the disappearance of this fossil in any stream be an event h. Similarly, the geologist notes the disappearance of a second fossil at levels that again differ from one stream to the next. Let this disappearance be an event *t*. Again, because of various 'accidents' (imperfections of the stratigraphical record) *h* and *t* are not always present in the same stream section, but do occur together in 14 of the stream beds. In 10 of these, *h* is stratigraphically above *t* while in the remaining 4, *h* is stratigraphically below *t*. Because of the *apparent* imbalance (10 to 4), the geologist may be persuaded to conclude that *h* and *t* are unrelated events and that *h* is really the younger. However, an analogy with the coin-tossing example above is worthwhile. If *h* and *t* occurred at the same time, then (*h* above *t*) should occur in the same frequency, approximately, as (*t* above *h*), any small deviation from equality being due to random fluctuations in preservation, exposure, etc. On the other hand, since (*h* above *t*) outnumbers (*t* above *h*) in the ratio 10:4, an alternative possibility is that *h* happened at a different time from *t* and is stratigraphically younger. The transfer to the six-step hypothesis-testing procedure is now immediate:

(1) H_0: (*h* above *t*) and (*t* above *h*) are equally likely because the two stratigraphical events are simultaneous. Any deviations from equality are due to sampling fluctuations.
    H_1: (*h* above *t*) outnumbers (*t* above *h*) because the two stratigraphical events are separate, *h* being the younger. (Note the alternative hypothesis has a 'direction' – namely 'younger'.)
(2) In any particular section, there are only two possible arrangements, (*h* above *t*) and (*t* above *h*) so that a test based on the binomial distribution is appropriate. Note the occurrence of *h* and *t* at the same level is excluded (corresponding to coins resting on their edges). If *h* and *t* occur at the same level in a very small number of streams, they are ignored. If they occured together in a very large number of streams, the approach would be quite different: the methods of Chapter 4 being applicable.
(3) $N = 14$, choose $a = 0.05$.
(4) The test statistic *R* is the number of occurrences of (*h* above *t*). It is binomially distributed so associated probabilities can be calculated.
(5) The one-tail critical region for $a = 0.05$ contains values of *R* of 11 or more (cf. *very many* heads).
(6) The observed value, $R = 10$, is not in the critical region. Therefore accept the null hypothesis.

Table 2.6 One-tail critical values of the test statistic R. Critical region includes values of R smaller than the critical value.

N	$a = 0.05$	$a = 0.01$	N	$a = 0.05$	$a = 0.01$	N	$a = 0.05$	$a = 0.01$
7	0	0	15	3	2	23	7	5
8	1	0	16	4	2	24	7	5
9	1	0	17	4	3	25	7	6
10	1	0	18	5	3	26	8	6
11	2	1	19	5	4	27	8	7
12	2	1	20	5	4	28	9	7
13	3	1	21	6	4	29	9	7
14	3	2	22	6	5	30	10	8

Because of the usefulness of this 'binomial test', one-tail critical values of the test statistic R for various values of sample size N and the two sizes of the critical region, $a = 0.05$ and $a = 0.01$, are given in Table 2.6. The table covers sample sizes up to $N = 30$. It should not be a difficult exercise to extend it, if needed.

3 One- and two-sample testing

In the application of simple statistical methods to our problems, the requirement often falls into one of two categories: we wish to answer one of two sorts of questions. Either 'Does this set of observations fit my theoretical predictions?' or 'Do these two sets of observations differ so much that I must assume they come from different populations?' These are 'one-sample' and 'two-sample' applications respectively, and I illustrate their structure by introducing the rapidly applied Kolmogorov–Smirnov tests, which operate directly on cumulative distribution functions (cdf's).

3.1 One- and two-sample tests

The last example in the previous chapter dealt with a case in which an observed situation (the number of superpositions of (*h* over *t*) is different from the number of superpositions of (*t* over *h*)) was compared with a theoretical prediction ((*h* over *t*) and (*t* over *h*) equally frequent). Another type of case frequently arising in practice involves the comparison of two observed situations with each other, in order to decide whether they are 'similar' or 'different'. The former case involves a one-sample test whereas the latter case requires a two-sample test. In short, a one-sample test comprises testing a randomly collected sample against a theoretical prediction for 'goodness of fit' between observation and expectation. On the other hand, a two-sample test is used to discover whether or not two independent samples could have come from the same population (or identical populations) and thus may be thought of as testing for comparability of two separate samples. To illustrate these ideas, the Kolmogorov–Smirnov tests are introduced, first to test the suggestion that 'basalt columns are six-sided' and then to investigate whether the range of cross-sectional shapes of basalt columns is very different from the range of shapes of mud flakes produced by desiccation cracks.

3.2 Polygonal shapes of basalt columns and mud flakes

Both mud flakes and cross sections of basalt columns approximate to polygons and we can 'measure' these polygons (according to our understanding of measurement in Ch. 1) by counting the number of their sides. Table 3.1 gives the results of such an operation amongst some naturally occurring examples.

We may have some interest in the mechanism of formation of these polygons. Popular theories suggest that they are due to shrinkage towards equidistantly spaced centres and should be six-sided, like the cells of a honeycomb. We note that the modal class amongst the basalt columns is certainly the six-sided one (almost half the sample is six-sided) but that there is some measure of spread, especially towards five- and four-sided columns, and that the total range goes from three to ten. Suppose we wished to test the popular theory that basalt columns are six-sided. How can we bring together observed fact and theoretical prediction? Some kind of graph is an obvious starting point and the data in this case can be expressed as a histogram. For the basalt columns, the summit of the histogram would correspond to six sides and the heights of the boxes would tail off in both directions. For the theoretical prediction, namely all basalt columns are six-sided, the histogram would have one box only, for the class of six-sided columns and the height of this box would be proportional to 33, the sample size. Unfortunately, we have no statistical test that can be based directly on such diagrams for such small sample sizes. In practice, a cumulative proportion 'curve' (or cumulative distribution function, Fig. 1.3b) would be much more useful and is shown in Figure 3.1.

Table 3.1 Polygonal shapes of basalt columns from Yellowcraigs, East Lothian, Scotland and mud flakes from New Haven, Connecticut, USA.

Classes (number of sides)	*Class frequencies*	
	Basalt	*Mud*
3	*1*	*1*
4	*3*	*7*
5	*8*	*10*
6	*15*	*8*
7	*4*	*6*
8	*1*	*4*
9	*0*	*0*
10	*1*	*0*
sample sizes	*33*	*36*

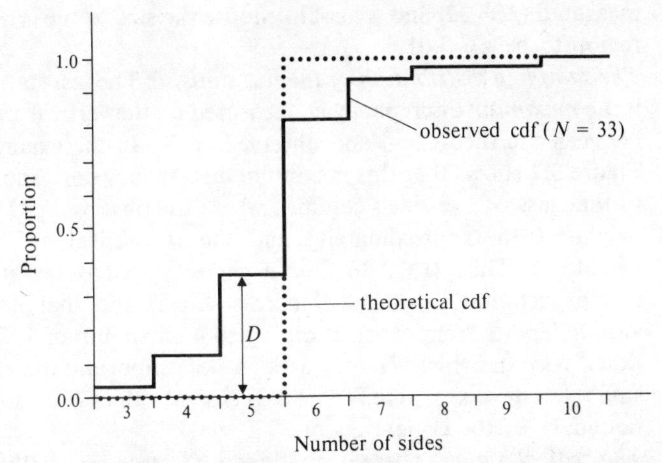

Figure 3.1 Theoretical and observed cumulative distribution functions for polygonal basalt columns thought to be six-sided.

As we should expect, there is a discrepancy between the theoretical and observed cdf's, and this discrepancy may be used as a measure of departure between expectation and observation. Our theoretical prediction is that shrinkage centres are equidistant and that resulting columns are six-sided. Our observations suggest that shrinkage centres are not perfectly equidistant so that we see columns that are not six-sided. But is the discrepancy significant or is it simply due to 'small' random fluctuations? We can use the six steps of hypothesis testing and the Kolmogorov–Smirnov one-sample test as follows.

3.3 The Kolmogorov–Smirnov 'one-sample' test

(1) *The null and alternative hypotheses.* The null hypothesis is that shrinkage centres are equidistant and that all resulting columns are six-sided. The alternative hypothesis is that shrinkage centres are not equidistant, they are distributed in some other way, not specified here, and the resulting columns show a range of polygonal shapes.

(2) *We choose the Kolmogorov–Smirnov one-sample test.* The test may be based on any kind of cumulative distribution function (thus any scale of measurement) and is not limited in its applicability by smallness of sample size.

(3) *Sample size and small probability a.* 33 columns have been

measured ($N = 33$) and we could choose the size of the critical region to be $a = 0.01$.

(4) *The sampling distribution of the test statistic.* The test statistic is the *maximum* discrepancy D, measured on the vertical scale, between the theoretical and observed cdf's. In the example, Figure 3.1 shows that this maximum discrepancy corresponds to the class of five-sided columns where the observed cdf has height $= 0.36$ (approximately) and the theoretical cdf has height $= 0$. Thus $D = 0.36$. For a perfect fit of observation with expectation, $D = 0$ and D increases the further that observation departs from expectation, up to a maximum of 1. The exact way in which D increases is not important here; it suffices to have available (Table 3.2) critical values of D at the boundary of the critical region.

(5) *The critical region.* There is no 'direction' associated with the alternative hypothesis in the sense of Section 2.5, because it is not important whether the observed distribution has more columns with fewer than six sides, or more columns with greater than six sides. (In this example, twelve columns have fewer than six sides, six columns have greater than six sides.) Consequently, a two-tail critical region is appropriate and referring to Table 3.2, we note that the critical region contains values of D of 0.28 and above.

(6) *The decision.* Our observed value of the test statistic, $D = 0.36$, falls in the critical region and thus we decide to reject the null hypothesis and accept that shrinkage centres are not spaced equidistantly.

Should we have chosen to frame an alternative hypothesis incorporating a sense of 'direction', then our means of calculating the value of the test statistic would need refinement, because discrepancies between the observed and the theoretical cdf's are also 'directional': they can be both *positive* (observed above theoretical) and *negative* (observed beneath theoretical). The particular discrepancy that would have to be taken as the value of the test statistic D would have to be consistent in its 'direction' with that of the alternative hypothesis (i.e. maximum positive *or* maximum negative).

Exercise (not easy). To consolidate your understanding of the last paragraph, test the null hypothesis, as above, against the alternatives that, 'on average', shrinkage centres are spaced such that columns have greater than* or fewer than* six sides (* delete as appropriate).

3.4 The Kolmogorov–Smirnov 'two-sample' test

The shapes of basalt columns and mud flakes can be compared graphically by preparing cumulative distribution functions, as in Figure 3.2. This is of interest if comparison is to be made between the respective mechanisms of formation, or indeed, between one of the mechanisms of formation (e.g. of basalt columns) in different geological settings. The prime interest now is whether or not the two samples could have come from the same (or identical) populations. The Kolmogorov–Smirnov two-sample test is appropriate, as follows.

(1) *The null and alternative hypotheses.* The null hypothesis is that the two samples are drawn from identical populations, as far as polygon shape is concerned, and that discrepancies are due to fluctuations of sampling. The alternative hypothesis is that the samples are drawn from different populations, without specifying *in any way* how these populations differ.
(2) The Kolmogorov–Smirnov two-sample test is appropriate because it can be based on any kind of cumulative distribution function. It is *independent* of any assumptions about the

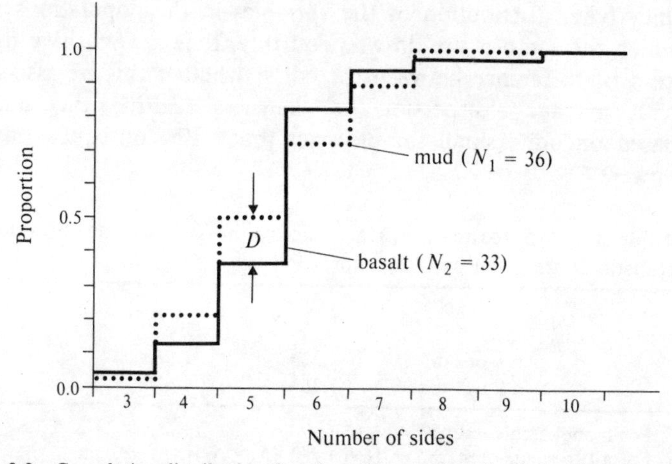

Figure 3.2 Cumulative distribution functions for samples of polygonal shapes of basalt columns and mud flakes laid out in preparation for the Kolmogorov–Smirnov two-sample test.

underlying forms of the *population distributions*, and it may be based on small sample sizes.

(3) The sample sizes are $N_1 = 36$ and $N_2 = 33$ and we choose a critical region of size 0.05.

(4) The test statistic has definition and properties very similar to those associated with the one-sample test above. In this case, since the alternative hypothesis does not imply any 'direction', the test statistic is the maximum discrepancy (irrespective of its 'direction') measured on the vertical scale between the two cdf's. It corresponds to the class of five-sided polygons and is given by (mud proportion) – (basalt proportion) = 0.50 – 0.36 = 0.14.

(5) *The critical region.* As this is a two-tail application (alternative hypothesis 'non-directional'), Table 3.2 gives the critical region of size 0.05 as containing all values of D of 0.33 and above.

(6) *The decision.* As our observed value of $D = 0.14$ does not fall into the critical region we do not reject the null hypothesis.

By way of conclusion, note the following features of the Kolmogorov–Smirnov tests. First, they are based on discrepancies between either an observed and a theoretical cdf or between two observed cdf's. Secondly, they may be used for variables on any scale of measurement: nominal, ordinal, ratio, continuous or discontinuous. Thirdly, no assumptions are made concerning the underlying distribution of the variables in the populations from which the samples are drawn. Fourthly, they are sensitive to any kind of difference between the cdf's, whether this be associated with 'average', 'dispersion' or 'skewness'. Fifthly, they may be based on quite small sample sizes (but with consequent loss of power).

Table 3.2 Approximate critical values of the Kolmogorov–Smirnov test statistic D are given by these expressions.

	$a = 0.05$	$a = 0.01$
one-tail	$1.22N'$	$1.51N'$
two-tail	$1.36N'$	$1.63N'$

For a one-sample test, $N' = 1/\surd(N)$.
For a two-sample test, $N' = \surd[(N_1 + N_2)/(N_1 \times N_2)]$.
The critical region contains values of D larger than the critical value. Sample sizes should be greater than about 16 for the above approximations to hold to about two significant figures.

4 Nominal-scale statistics

The compilation of a basic description is fundamental to the observational sciences, soon followed by a classification of the objects or events of interest and then an analysis of the compositions of populations (of the statistical type) based on counts of the frequencies of individuals falling into each class. Here we look at some methods of analysing data that consist of these frequencies of occurrence, numbers obtained simply by counting how many individuals of a particular type our sample contains. Many classifications consist of setting up a number of points on a nominal scale, one of the simplest being to classify objects according to whether they show a specific characteristic or do not show that characteristic.

Such frequency data is conveniently presented in the form of tables and certain properties of such 'contingency tables' are readily analysed. The tests here are related to an example in which porphyroblasts of different mineral species occur in gneisses. On the way through, we have to look at a new statistical concept, that of 'degrees of freedom', but I hope we shall not side-track too much.

4.1 Nominal scale – further developments

We have seen in Chapter 1, during our brief discussion of measurement theory, that the rules of rock, mineral and fossil identification may be treated as means of placing objects in their appropriate positions on a nominal scale of measurement. Moreover, individual objects may be measured on more than one nominal scale simultaneously. For example, rock fragments from a glacial deposit may be classified according to (a) rock type and (b) location of collection site. The resulting frequencies can then be analysed to discover whether there are significant differences in the balance of rock types between sites. To illustrate the methods, however, another slightly different method of using the nominal scale is introduced.

Geological objects can frequently be classified simply according to the presence or absence of a particular attribute (i.e. on a two-point nominal scale). The numerical example I give below is based

Table 4.1 Contingency table for porphyroblastic localities.

	H	*h*	*Row totals*
B	(15)	(18)	(33)
b	(48)	(9)	(57)
Column totals	(63)	(27)	$N = 90$

on a sample of 90 localities in a formation of porphyroblastic gneiss in south Skye, Scotland. At each locality, one or more of the three mineral species: hornblende, biotite and garnet is present as porphyroblasts set in a quartzofeldspathic matrix. Each locality is therefore measurable on three two-point nominal scales, namely hornblende (present or absent), biotite (present or absent), and garnet (present or absent). For this example, I want to concentrate on two of these, hornblende and biotite. If *H* denotes the presence of hornblende and *h* denotes its absence, and similarly *B* and *b* for the presence and absence of biotite, then individual localities can be labelled *HB*, *Hb*, *hB* or *hb* according to which porphyroblasts are present. The total number (or frequency) of localities characterised by a particular porphyroblast association can then be denoted by the appropriate symbols between round brackets: (*HB*), (*Hb*), etc. In this numerical example, $(HB) = 15$, $(Hb) = 48$, $(hB) = 18$, $(hb) = 9$, total $N = 90$. As we have only two nominal scales in use, we can present these figures as in Table 4.1, called a **contingency table**.

4.2 Contingency tables – independence

As Table 4.1 has only two rows and two columns, it is known as a 2×2 contingency table. Having measured two variables and displayed their frequencies in this contingency table, the prime question is: 'Are these two variables independent?' In other words, is the occurrence or otherwise of biotite porphyroblasts in these rocks totally unconnected (i.e. independent) of the occurrence or otherwise of hornblende porphyroblasts? Both are ferromagnesian minerals of broadly similar chemical composition and whether one or the other or both tend to occur in rocks of a given composition is largely determined by prevailing metamorphic conditions. If statistical analysis leads us to conclude that the hornblende and biotite porphyroblasts occur independently, then one particular set of metamorphic conditions is implied, if we conclude that they do not occur independently, then another set of metamorphic conditions is implied. But what is our criterion of independence?

First, let us put symbols into the contingency table (Table 4.2). Now the criterion of independence can be stated in two forms:

(a) The proportion of biotite rocks that contain hornblende (i.e. $(HB)/(B)$) should be the same as the proportion of non-biotite rocks that contains hornblende (i.e. $(Hb)/(b)$). (Compare these two ratios with the positions of their component variables in Table 4.2.)
(b) The proportion of biotite rocks that contains hornblende (i.e. $(HB)/(B)$)should be the same as the proportion of all rocks that contains hornblende (i.e. $(H)/(N)$). (As above, compare these two ratios with the positions of their component variables in Table 4.2. Do you see a pattern emerging?)

Even though these two variables may be truly independent, the equalities implied by these two forms of the criterion will rarely be perfectly balanced because of fluctuations in the sampling of the population. If we assume independence as a null hypothesis, then we can calculate the expected values of the frequencies. Enclosing expected frequencies between double quotes (e.g. "*HB*"), we can use the second form of the criterion to derive:

$$\text{“}HB\text{”} = (B) \times (H)/N \tag{4.1}$$

Exercise. Prepare a contingency table containing the numerical values of the *expected* class frequencies. Do this simple exercise now – you will need the answers to it below.

Table 4.2 Contingency table with frequencies denoted by symbols (symbols in round brackets are the observed frequencies).

	H	*h*	*Row totals*
B	(*HB*)	(*hB*)	(*B*)
b	(*Hb*)	(*hb*)	(*b*)
Column totals	(*H*)	(*h*)	*N*

Table 4.3 Contingency table showing discrepancies of observed and expected frequencies. The observed frequencies are expressed in terms of the expected frequencies and the discrepancy *d*.

	H	*h*	
B	"*HB*" + *d*	"*hB*" − *d*	(*B*)
b	"*Hb*" − *d*	"*hb*" + *d*	(*b*)
	(*H*)	(*h*)	*N*

4.3 Positive and negative association

In our example "*HB*" = 23, which is rather greater than the observed frequency (*HB*) = 15. If the discrepancy did not arise due to sampling fluctuation in a population in which the variables are truly independent, then it must indicate an 'association' between the variables – in this case a 'negative' association since we have fewer hornblende–biotite rocks than expected. In other words the rocks tend to be either biotite-bearing or hornblende-bearing but not both. The opposite, with a tendency for hornblende and biotite to occur together (thus (*HB*) > "*HB*") would illustrate 'positive' association. If the discrepancy is denoted by $d = (HB) - \text{"}HB\text{"}$, then it is easy to show, with row and column totals held fixed, that the contingency table may be rewritten, symbolically, to become Table 4.3.

4.4 The 'chi-squared' test

With $d = 15 - 23 = -8$, the possibility has been suggested that this discrepancy arises simply due to sampling fluctuation. A statistical test is needed to assess the probability of occurrence of a value of *d* of this magnitude. The chi-squared test is suitable and I introduce it here via the six steps of hypothesis testing:

(1) *State the null and alternative hypotheses.* The null hypothesis is that hornblende and biotite are of independent occurrence, but because of sampling fluctuation, (*HB*) is 15 instead of its expected value of 23. The alternative hypothesis is that the discrepancy is real so that hornblende and biotite are negatively associated.

(2) *Choose the statistical test.* In general, the test statistic χ^2 (chi-squared) is a measure of the departure of observation from expectation in the case of frequencies that arise from *any* classification of data in which each specimen is put into one, and only one, class. In the example the data have been allotted to four classes in a 2×2 contingency table. Limitations on the chi-squared test arising theoretically restrict its application to samples greater than about 50 and cases in which the lowest expected frequency in any class is not less than five, preferably ten or more. Here, the sample size is 90 and the lowest expected frequency "hB" = 10, so the chi-squared test is applicable. (Should expected class frequencies be too small, then adjacent classes should be combined if possible and realistic, until the above condition is satisfied.) All classes should be independent of each other, i.e. a specimen should be allotted to one class only, not more. χ^2 may be calculated from either of the two expressions:

$$\chi^2 = \sum_{i=1}^{i=C} [(O_i - E_i)^2 / E_i] \tag{4.2}$$

$$= \sum_{i=1}^{i=C} (O_i^2 / E_i) - N \tag{4.3}$$

where C is the number of classes, i is an integer (i.e. a whole number) labelling the classes and taking values from 1 to C in steps of 1, O_i is the observed frequency in the ith class, E_i is the expected frequency in the ith class, N is the sample size and Σ denotes the operation of summation of the expression *within the outermost brackets* over all of the values of i specified (cf. Eq. 1.2 for an introduction to this usage of the summation sign).

If the classes in the 2×2 contingency table (Table 4.1) of the example are labelled $i = 1, 2, 3, 4$, clockwise, starting at the top left, then Equation 4.3 gives:

$$\chi^2 = \frac{15^2}{23} + \frac{18^2}{10} + \frac{9^2}{17} + \frac{48^2}{40} - 90 = 14.5$$

(Note that the summation is completed, *then* the sample size N is subtracted.) Of the two, Equation 4.3 is usually more easily applied to the calculation of χ^2.

(3) *Choose the sample size and the size of the critical region.* Here $N = 90$. Let the size of the critical region be $a = 0.01$.

(4) *Investigate the sampling distribution of the test statistic.* The test statistic χ^2 has the following properties:

   (a) It takes positive values only, increasing from zero but with no upper limit.

   (b) It increases as observation departs further from expectation.

   (c) Its precise value is dependent on the number of 'degrees of freedom' of the classification. In the case of the 2×2 contingency table, the number of degrees of freedom of the classification is one because only one class frequency can be chosen arbitrarily, subject to the marginal totals of the contingency table remaining fixed. I shall return briefly to the general idea of degrees of freedom of data sets in the following section. In the meantime, we note that critical values of χ^2 can be found in Table 4.6.

(5) *Define the critical region.* Table 4.6 gives a critical value of χ^2 of 6.64 for a critical region of size 0.01 and a classification having 1 degree of freedom. From Equation 4.2, we see that χ^2 increases as observation departs further from expectation, so the critical region contains values of χ^2 of 6.64 and greater.

(6) *The decision.* The observed value of $\chi^2 = 14.5$ lies within the critical region so the null hypothesis is rejected and we conclude that the hornblende and biotite porphyroblasts are truly negatively associated. There is a probability of $a = 0.01$ or less that we are mistaken in this conclusion.

(Subsequent microscope work indicated that a replacement relationship existed between the hornblende and the biotite.)

4.5 Degrees of freedom

I have mentioned this topic already in the preceding section and shall have to do so again later in the book, so let me try to outline the concept as follows. When we process the data from a sample so as to produce a classification or to calculate a number (or a set of numbers), including both parameters and test statistics, then the original data plus the classification (or the derived numbers) are clearly linked by some overall structure. In the case of a classification, this structure consists of the classification rules, in the case of the derived numbers, it is summarised by one or more mathematical expressions. The number of degrees of freedom of the structure is defined as the sample size (or the number of classes, as appropriate) less the number of independent constraints imposed by the structure. It tells us something about the number of ways the sample may be changed without having to produce any change in the constraining factors.

Let me try to illustrate this with two examples. In the last section, we looked at a 2×2 contigency table in a search for independence of attributes. The constraints applying to the table are that we can vary the frequency in any class subject to the marginal totals (which we use to define independence of attributes) remaining fixed. There are four marginal totals, thus four constraints, namely two column totals and two row totals, and they sum separately to the overall sample size. However, the latter is clearly not an independent constraint. Further, as the two column totals and the two row totals must sum to the same number, all four are not independent: if any three are specified, then the fourth is automatically fixed. Thus only three of the constraints are *independent* and the number of degrees of freedom of the 2×2 contingency table with fixed marginal totals is:

$$4 \text{ classes} - 3 \text{ independent constraints} = 1$$

In Section 4.4 above, we saw indeed that only one class frequency in the 2×2 table could be chosen arbitrarily subject to fixed marginal totals. Once one class frequency is determined, all three others are then fixed.

For the second example, take the case of the calculation of the mean value of a variable in a sample of size N and then answer the question: 'How many of the individual specimen values may be altered arbitrarily subject to the mean value remaining fixed?' The answer is that we can assign any values we care to to the first $N-1$ specimens but the value for the Nth is constrained by the fact that the mean value is fixed. Thus an estimate of a mean value is said to have $N-1$ degrees of freedom. Had we also estimated the standard deviation, then the overall structure would have $N-2$ degrees of freedom because these two parameters are independent. So long as the parameters and test statistics we evaluate are independent, then we simply subtract their number from the sample size to discover the number of degrees of freedom.

4.6 The Fisher exact probability test

We noted above that the chi-squared test can be used only on samples above a certain size, with individual class frequencies also above a certain minimum. Indeed, in the example, the latter condition was only just satisfied. In circumstances where the chi-squared test becomes inapplicable to 2×2 contingency tables because of these factors, we can usually substitute the Fisher exact probability test. Thus, at an earlier stage in the study described above, the contingencies shown in Table 4.4 were found. To illustrate the use of the Fisher exact probability test, the six steps of hypothesis testing may be used again:

(1) *State the null and alternative hypotheses.* These are worded as before except that (HB) is 3 instead of its expected value of 7.
(2) *Choose the statistical test.* Because the sample size is less than 50 and three of the expected class frequencies are less than 10, the chi-squared test is not applicable and the Fisher exact probability test is to be used instead.
(3) *Choose the size of sample and the size of the critical region.* Sample size $N = 30$. Because of this, the test is likely to be less powerful (in the sense used in Section 2.4) so a suitable size for the critical region might be $a = 0.05$.
(4) *Study the sampling distribution of the test statistic.* There is no explicit test statistic associated with the Fisher exact probability test, instead we go from the contingency table directly to its probability of occurrence, as shown below. The observed frequencies do not coincide with the expected frequencies *predicted by the null hypothesis* but this could be due to sampling fluctuation. The probability of getting the observed frequencies from the expected frequencies, given sampling

Table 4.4 Contingency table arising at an earlier stage in the study of porphyroblastic gneisses (round brackets contain observed frequencies, double quotes enclose expected frequencies).

	H	h	
B	(3) "7"	(6) "2"	(9)
b	(19) "15"	(2) "6"	(21)
	(22)	(8)	30

fluctuation, is given exactly by:

$$p_0 = \frac{(B)!\,(b)!\,(H)!\,(h)!}{N!\,(HB)!\,(Hb)!\,(hB)!\,(hb)!} \tag{4.4}$$

where '!' is the factorial symbol (Eq. 2.1). This is readily evaluated using Table 4.7.

$$p_0 = \frac{9!\,21!\,22!\,8!}{30!\,3!\,19!\,6!\,2!} = 0.003014$$

But, under the null hypothesis true, more extreme values of observed frequencies could have occurred also (i.e. observed frequencies even further away from the expected frequencies). Thus the smallest observed frequency in Table 4.4 is $(hb) = 2$, with a probability of occurrence of 0.003014, as calculated above. But to this we must add the probabilities of occurrence of the even smaller (or more extreme) frequencies $(hb) = 1$ and $(hb) = 0$, given the *same marginal totals.* These more extreme contingency tables and their probabilities of occurrence, as calculated from Equation 4.4, are given in Table 4.5, from which, the probability of occurrence of the observed table, or one more extreme, is the sum of these individual probabilities:

$$p = p_0 + p_i + p_{ii} = 0.003145 \text{ (or 0.003 for practical purposes)}$$

Table 4.5 More-extreme contingency tables. (The asterisk marks the cell with lowest frequency.)

	H	*h*		
B	(2)	(7)	(9)	
b	(20)	(1)*	(21)	$p_i = 0.000129$
	(22)	(8)	30	

	H	*h*		
B	(1)	(8)	(9)	
b	(21)	(0)*	(21)	$p_{ii} = 0.000002$
	(22)	(8)	30	

(5) *Define the region of rejection.* The size of the critical region is $a = 0.05$ and thus our observed frequencies will fall into the region if the probability of their occurrence *plus* the probability of occurrence of more extreme frequencies is, in total, *smaller* than this region.

(6) *The decision.* The probability of occurrence of the observed frequencies, plus those more extreme, is less than 0.05. The null hypothesis is rejected and we conclude as before, that hornblende and biotite are negatively associated. The probability of our being mistaken in this conclusion is 0.003 or less.

4.7 Discussion

Two matters remain for comment. First, it is useful to note how the power of the tests increases with the sample size. At an early stage of the study described above, when only 30 porphyroblastic localities had been visited, the null hypothesis of hornblende and biotite independence could be rejected with a probability of error of 0.05 or less (0.003 or less, to be exact). However, when the sample size is increased to 90, the null hypothesis could then be rejected with a probability of error of 0.01 or less (actually 0.0000001 is nearer the mark).

Secondly, the example has confined discussion to 2×2 contingency tables. More-refined studies would involve larger p by q tables or even p by q by r – three-dimensional contingency tables. The

Table 4.6 Critical values of χ^2 for critical regions of size $a = 0.05$ and $a = 0.01$ and for classifications of various degrees of freedom ν. The critical regions contain values of χ^2 greater than the critical values.

ν	$a = 0.05$	$a = 0.01$	ν	$a = 0.05$	$a = 0.01$	ν	$a = 0.05$	$a = 0.01$
1	3.841	6.635	13	22.36	27.69	25	37.65	44.31
2	5.991	9.210	14	23.68	29.14	26	38.89	45.64
3	7.816	11.35	15	25.00	30.58	27	40.11	46.96
4	9.488	13.28	16	26.30	32.00	28	41.34	48.28
5	11.07	15.08	17	27.59	33.41	29	42.56	49.59
6	12.59	16.81	18	28.87	34.81	30	43.77	50.89
7	14.07	18.49	19	30.14	36.19	40	55.76	63.69
8	15.51	20.09	20	31.41	37.57	50	67.50	76.15
9	16.92	21.67	21	32.67	38.93	60	79.08	88.38
10	18.31	23.21	22	33.92	40.29	70	90.53	100.4
11	19.68	24.72	23	35.17	41.64	80	101.9	112.3
12	21.03	26.22	24	36.42	42.98	90	113.1	124.1

Table 4.7 Logarithms (to base 10) of the factorials of numbers up to 50: of use in the calculation of probabilities in the Fisher exact probability test.

n	$\log_{10}n!$	n	$\log_{10}n!$	n	$\log_{10}n!$	n	$\log_{10}n!$	n	$\log_{10}n!$
1	0.0000	11	7.6012	21	19.7083	31	33.9150	41	49.5244
2	0.3010	12	8.6803	22	21.0508	32	35.4202	42	51.1477
3	0.7782	13	9.7943	23	22.4125	33	36.9387	43	52.7811
4	1.3802	14	10.9404	24	23.7927	34	38.4702	44	54.4246
5	2.0792	15	12.1165	25	25.1906	35	40.0142	45	56.0778
6	2.8573	16	13.3206	26	26.6056	36	41.5705	46	57.7406
7	3.7024	17	14.5511	27	28.0370	37	43.1387	47	59.4127
8	4.6055	18	15.8063	28	29.4841	38	44.7185	48	61.0939
9	5.5598	19	17.0851	29	30.9465	39	46.3096	49	62.7841
10	6.5598	20	18.3861	30	32.4237	40	47.9116	50	64.4831

former are easily constructed and manipulated by the methods using the chi-squared test, described above. The more complicated three-dimensional developments are beyond the scope of this book but the interested reader is referred to Yule and Kendall (1950). Significant values of χ^2 arising from p by q tables may require careful analysis of the data to locate the source of the significance. In the case of p by q by r tables, associations between the variables may or may not be real, apparently significant associations between two of the variables arising because both are, in reality, associated independently with the third. Nevertheless, the methods are of general interest and are worthy of careful consideration in a variety of situations.

5 Theoretical distributions and confidence intervals

So far, we have not worried too much about exactly how our variables are distributed in the population, i.e. what the pdf's and cdf's look like, beyond discovering that the parameters mean, standard deviation and skewness can tell us something about the position and shape of the distribution. It would be neat and tidy if Nature provide populations that followed exact mathematical rules, but she does not, we have to make approximations or assumptions in our clumsy way. The usual justification for doing this is so that further mathematical manipulation of the data becomes easier. So long as we keep a careful eye on these assumptions, we can do a number of useful things, like attaching a number to our level of confidence in an estimate of a parameter. This chapter starts with some outline theoretical work and shows how confidence can be quantified. Then follows yet another concept by which statistical tests can be classified into parametric and non-parametric types, accompanied by some discussion of their relative merits. Lastly, a method of placing confidence limits on proportions estimated by counting is given, applicable for example to quantifying the content and ratio of olivine and augite phenocrysts in a lava.

5.1 The place of theoretically derived distributions

Theoretically derived probability density functions of variables ('distributions' for brevity, if the context is clear) are used frequently in the development of practical statistical methods. I have already touched upon one, the binomial distribution, in the discussion of coin tossing (with stratigraphical applications) in Chapter 2, and the further elaboration of statistical methods is impossible without them. One theoretical distribution, of central importance in the study of *ratio* scale variables, is the Gaussian or 'normal' distribution. This was derived in astronomy during the development of the theory of errors of measurement of star positions, but many naturally occuring distributions in the observational sciences seem to be closely approximated by the Gaussian distribution: thus its alternative name: 'normal'. Indeed, because of this good approximation, many of the more powerful statistical testing procedures are based on the assumption that the underlying variable has a Gaussian distribution, so below we shall derive methods of testing this basic assumption. Furthermore, because the Gaussian distribution is easily manipulated mathematically, it features prominently in further refinements of statistical methods and we shall encounter it in this guise in later chapters.

The test statisitic Student's t is useful in a variety of applications in which the underlying *ratio* scale variable can be assumed to have a Gaussian distribution. Not only can Student's t distribution be used as a powerful method of testing for differences between samples but it also allows us to calculate 'confidence intervals' which tell us something about the reliance we may place on our estimates of population parameters such as the mean. I shall also introduce a method of calculating confidence intervals relating to *proportions* estimated by counting, e.g. as a result of point-counting minerals in thin sections, but also useful in similar field observations.

5.2 The Gaussian pdf

5.2.1 The curve

The probability density function of the Gaussian distribution is a symmetrical, single-humped curve with tails extending to infinity in both directions. It is shown in Figure 5.1, fitted to the frequency distribution of a sample of pebbles, the maximum diameters, or 'lengths', of which have been measured.

The pdf of the Gaussian distribution is given here, but *only* for reference, by:

$$f(X) = \exp\ [-(X-M)^2/(2s^2)]/\sqrt{2\pi s^2} \qquad (5.1)$$

where X is the variable. M and s are *parameters* (constants) of the Gaussian distribution called the 'mean' and 'standard deviation' respectively and exp (x) means exponential e raised to the power (x). The quantity s^2 is called the variance and is the square of the standard deviation.

5.2.2 Estimation of parameters

If we are confident that our sample comes from a parent population in which the variable of interest has a Gaussian distribution, then

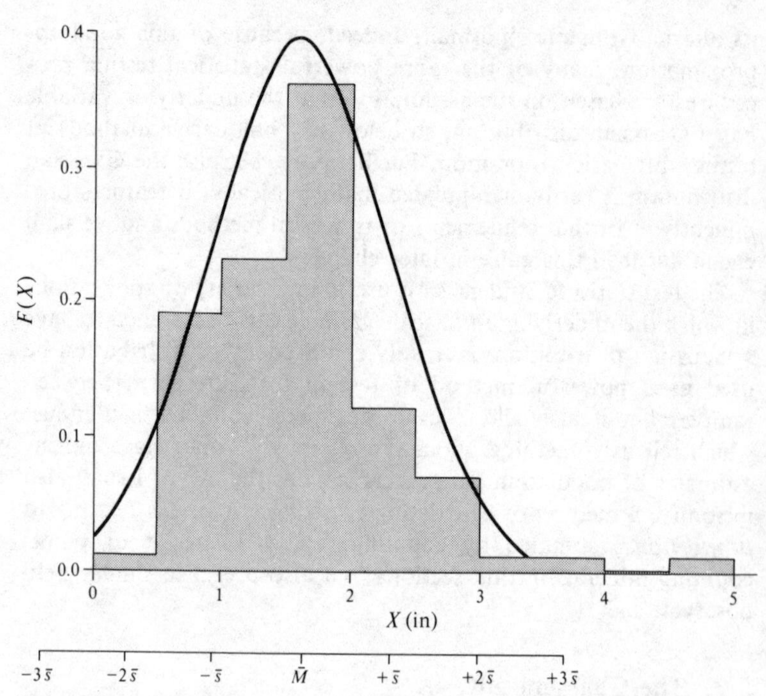

Figure 5.1 A Gaussian pdf fitted to a sample of beach pebbles. The histogram displays the data presented in Table 5.1. The smooth, single-humped continuous curve is the Gaussian pdf calculated so as to have the same mean $\overline{M}$ and the same standard deviation $\bar{s}$ as the pebble sample. The lower of the two horizontal scales shows the pebble maximum diameters as deviations, measured from the mean, with one standard deviation as the unit. Note that virtually the whole of the theoretical distribution is encompassed by the interval $\overline{M} \pm 3 \times \bar{s}$.

it is often convenient to replace all of the separate measurements of the variable in the sample by just two figures, namely our *estimates* of the two *parameters* of the parent population. These two estimates are of the mean $\overline{M}$ (where the bar above the symbol emphasises the fact that it *is* an estimate of the *true*, but unknown, value of the mean M), and the standard deviation $\bar{s}$, the square root of the variance. The **mean** is the value of the variable about which the sample is clustered, the **standard deviation** is a measure of the amount of 'spread' or 'dispersion' of the sample cluster. Note that the standard deviation has the same units of measurement as the mean (e.g. mm or kg or whatever) and that, as you will be able to verify presently, a proportion of 0.68 (about two thirds) of the sample values are contained in the range of measurements from (mean minus standard deviation) to (mean plus standard deviation). Further, as you see from Figure 5.1, almost the whole of the theoretical Gaussian distribution (a proportion of 0.997, to be more precise) is contained in the interval; mean plus or minus three times the standard deviation.

If we have a sample that we suspect is drawn from a population that follows a Gaussian distribution, then we have to *estimate* the population parameters M and s. This may be done using the following expressions.

The mean (or 'average') is *estimated by:*

$$M = \frac{1}{N} \sum_{i=1}^{i=N} (X_i) \tag{5.2}$$

and the standard deviation (a measure of 'dispersion' about the mean) is *estimated* by:

$$\bar{s}_2 = \frac{1}{N-1} \sum_{i=1}^{i=N} (X_i - \overline{M})^2 \tag{5.3a}$$

$$= \frac{1}{N-1} \sum_{i=1}^{i=N} (x_i^2) \tag{5.3b}$$

$$= \frac{1}{N-1} \sum_{i=1}^{i=N} (X_i^2) - \overline{M}^2 \tag{5.3c}$$

where N is the sample size, i labels specimens from 1 to N, X_i is the value of the variable associated with the ith specimen, Σ means summation of the expression in the outermost brackets over the values of i specified and $x_i (= X_i - \overline{M})$ is the **deviate** measured from the mean.

Unless a calculator has specially incorporated functions, Equation 5.3c is probably the simplest by which the standard deviation may be estimated, but be careful of numerical imprecision as discussed in Appendix C. Should the data have been classified (or 'grouped' – Section 1.3.1), then the mean and standard deviation are estimated by:

$$\overline{M} = \frac{1}{N} \sum_{i=1}^{i=C} (f_i X_i) \tag{5.4}$$

$$\bar{s}^2 = \frac{1}{N-1} \sum_{i=1}^{i=C} (f_i X_i^2) - \overline{M}^2 - \frac{w^2}{12} \tag{5.5}$$

where C is the number of classes, i labels the classes from 1 to C, f_i is the frequency in the ith class, X_i is the value of the variable at the mid-point of the ith class and w is the class width.

Table 5.1 Maximum diameters (X in) in a sample of 134 pebbles.

Classes		*Class frequencies*	*Cumulative frequencies*	*Cumulative proportions*
>0 in	⩽0.5 in	1	1	0.007
0.5	1	25	26	0.194
1	1.5	31	57	0.425
1.5	2	48	105	0.784
2	2.5	16	121	0.903
2.5	3	10	131	0.978
3	3.5	1	132	0.985
3.5	4	1	133	0.993
4	4.5	0	133	0.993
4.5	5	1	134	1.000

Exercise. In the case of the pebble diameters of Figure 5.1, verify from the data in Table 5.1, using Equations 5.4 and 5.5, that the mean diameter is 1.62 in and the standard deviation is 0.68 in.

5.3 Testing for Gaussian distribution of an observed variable

Because several of the more powerful statistical tests depend on the assumption that the underlying variable (or variables) has (or have) a Gaussian distribution, it is important in such cases to test sample data against a Gaussian form for 'goodness of fit'. Although a number of relatively informal methods have been suggested to do this, we are likely to find the Kolmogorov–Smirnov one-sample test particularly useful and rapid. To apply this test, we compare an observed cumulative distribution function with a predicted cumulative distribution function, the latter calculated so as to have M and s the same as those estimated from the sample.

To illustrate the application of such a test, we may follow the six steps of the hypothesis-testing procedure, using the maximum diameters of the pebble sample plotted in Figure 5.1 as an example:

(1) *State the null and alternative hypotheses.* The null hypothesis states that maximum diameters of the pebbles are drawn from a Gaussian population which has mean maximum diameter 1.62 in with standard deviation 0.68 in. The alternative hypothesis states that the pebble maximum diameters are drawn from a population which is not as specified in the null hypothesis, but does not state *how* the population differs.

(2) *Choose the statistical test.* The Kolmogorov–Smirnov one-sample test (see Sections 3.2 & 3) is appropriate in that it compares an observed cdf with a predicted cdf. Measurements in this case are on continuous, ratio scales, so the cdf's are smooth curves. The Kolmogorov–Smirnov test makes no assumption about the distribution of the variable in question, outside the assertions of the null hypothesis above.

(3) *Choose a size for the sample and for the critical region.* The sample of pebbles comprises 134 specimens. Let the size of the critical region be 0.01.

(4) *Examine the sampling distribution of the test statistic.* The Kolmogorov–Smirnov test statistic is D, the maximum discrepancy between the two cdf's as measured on the vertical scale. It is zero for perfect fit and increases to a limit of 1 as observation departs from expectation. Critical values have been given in Table 3.2.

(5) *Define the critical region.* There is no implication of a 'direction' in the statement of the alternative hypothesis, so a two-tail test is being applied. Reference to (4) above and Table 3.2 shows that the critical region contains values of D of 0.14 and greater.

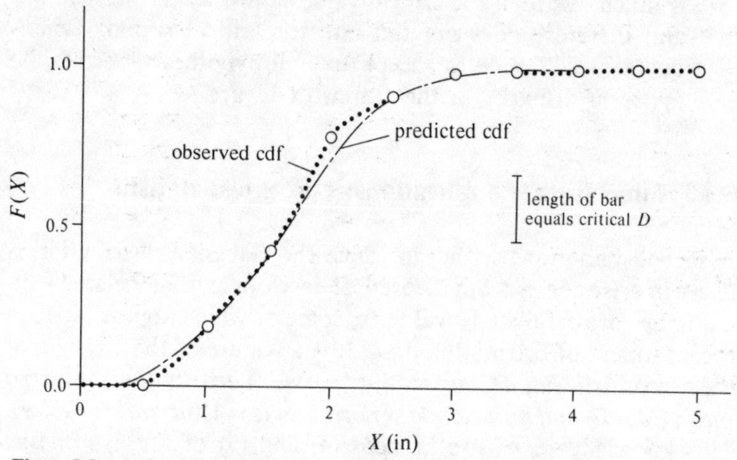

Figure 5.2 A Gaussian cdf fitted to a sample of beach pebbles. The dotted curve is constructed from the data of Table 5.1. The predicted Gaussian cdf is shown by the pecked line, and is calculated on the basis of mean and standard deviation estimated from the sample. Details are given in Table 5.2. The length of the vertical bar is equal to the critical value of the Kolmogorov–Smirnov test statistic D. Since the vertical bar cannot be fitted between the curves, the null hypothesis is not rejected.

Table 5.2 Calculation of variable values X for various values of the standardised deviate z (see Table 5.4) in a population with estimated mean $\overline{M} = 1.62$ in and estimated standard deviation $s = 0.68$ in.

z	X(in)	$F(X)$	z	X(in)	$F(X)$
−3.5	−0.76	0.000	+0.1	+1.69	0.540
−3.0	−0.42	0.001	+0.3	+1.82	0.618
−2.5	−0.08	0.006	+0.5	+1.96	0.692
−2.0	+0.26	0.023	+1.0	+2.30	0.841
−1.5	+0.60	0.067	+1.5	+2.64	0.933
−1.0	+0.94	0.159	+2.0	+2.98	0.977
−0.5	+1.28	0.309	+2.5	+3.32	0.994
−0.3	+1.42	0.382	+3.0	+3.66	0.999
−0.1	+1.55	0.460	+3.5	+4.00	1.000

(6) *The experiment and the decision.* The observed cumulative distribution function is constructed as outlined in Section 1.3.4, from the data in Table 5.1 and is incorporated in Figure 5.2. The predicted cumulative distribution function can be calculated from the standardised data of Table 5.4 (examine it now) via Table 5.2, and plotted in Figure 5.2. From the graphs of Figure 5.2, it is clear that the value of the test statistic D (the maximum vertical separation of the two cdf's) is much less than the critical value shown by the vertical bar and therefore does not fall into the critical region. Consequently, we decide to accept the null hypothesis, despite the apparent disparity of the graphs of Figure 5.1.

5.4 Some further applications: z as a test statistic

A second important setting in which the Gaussian distribution is likely to arise comes from theoretical developments associated with a number of useful statistical tests, some of which appear later, in the treatment of orientation data. It is a feature of these tests that the test statistic *has a Gaussian (or 'normal') distribution with zero mean* ($M = 0$) *and unit standard deviation* ($s = 1$)(*or unit variance*). This knowledge is of use in steps (4) and (5) of the hypothesis-testing procedure where the critical region of the test statistic is defined.

For such test statistics, with $M = 0$, a zero value indicates perfect agreement of observation with expectation in a one-sample case or of two samples in a two-sample case. The further that observation departs from expectation or the two samples depart from each

Table 5.3 Critical values of the standardised deviate.

	a = 0.05	*a* = 0.01
one-tail	1.64	2.33
two-tail	1.96	2.58

Table 5.4 The cumulative distribution function of the Gaussian distribution. For convenience of tabulation the cdf is given as a function of the **standardised deviate,** z, where z is given by: $z = (X - \overline{M})/\bar{s}$. To convert a standardised deviate to a variable value, this equation may be rearranged: $X = z \times \bar{s} + \overline{M}$.

Standardised deviate	*Cumulative proportion*	*Standardised deviate*	*Cumulative proportion*
−3.5	0.000	+0.1	0.540
−3.0	0.001	+0.3	0.618
−2.5	0.006	+0.5	0.692
−2.0	0.023	+1.0	0.841
−1.5	0.067	+1.5	0.933
−1.0	0.159	+2.0	0.977
−0.5	0.309	+2.5	0.994
−0.3	0.382	+3.0	0.999
−0.1	0.460	+3.5	1.000

other, so the value of the test statistic increases, either in the positive or negative direction. The critical regions, therefore, lie under one or both tails of the Gaussian pdf (Fig. 5.1) according to whether a one-tail or a two-tail test is being applied. All such tests include the application of an equation whereby we can calculate a value of the **standardised deviate** z (see Table 5.4) and use it as our test statistic. In order to define critical regions, therefore, a table of critical values of z is needed (Table 5.3).

5.5 Confidence intervals – introductory

In the example above, I have *estimated* the mean maximum diameter of a sample of 134 pebbles to be 1.62 in. Although this figure is quoted to the second decimal place, it is very unlikely that a second sample drawn from the same population would give the same answer. Indeed, a careful repetition of the experiment with a different sample of 90 pebbles produced an answer of 1.69 in (with standard deviation 0.62 inches). So can I quote a meaningful result?

This is a universal problem in the field of estimation and the standard solution is to state an *estimate* of a particular population parameter together with a 'confidence interval (or set or region)'. In the case above, the result is: the mean maximum diameter of a sample of 134 pebbles is estimated to be 1.62 in with a probability of 0.99 that the true value lies in the interval 1.47 to 1.77 in. Or rather more loosely, 'I am 99% confident that the mean maximum diameter is between $1\frac{1}{2}$ and $1\frac{3}{4}$ in'. Such confidence intervals are calculated as below.

5.6 Student's *t* and confidence intervals

In the context of the present example (i.e. estimation of means), the calculation of the confidence interval is based on a knowledge of Student's *t* distribution. This distribution arose during a study of the theory of errors of estimation in relation to small samples, where the classical methods became inadequate. The test statistic, Student's *t*, has a distribution graphically very similar to the *z* of the Gaussian distribution in that its pdf is single-humped and is symmetrical about $t = 0$. It differs in being associated with a number of degrees of freedom, ν, which is nearly always a function of sample size(s). However, once ν is determined, critical values of *t* may be obtained from tables (e.g. Table 5.5) and one-tail and two-tail critical regions may be defined, so we need pay little attention to the precise details of the pdf of *t*.

To show the calculation of a confidence interval associated with the pebble diameters, I shall try to adapt the six-step hypothesis-testing procedure although alert students will soon find several short cuts, since there is a distinct parallel with a one-sample, two-tail test.

(1) *State the null and alternative hypotheses.* The null hypothesis is that the sample of 134 pebbles is drawn from a population in which the maximum diameters have a Gaussian distribution with standard deviation 0.68 in and with mean somewhere *inside* the interval M_l to M_u, where these lower and upper bounds (respectively) remain to be calculated. The alternative hypothesis is similarly worded except that the mean lies somewhere *outside* the interval M_l to M_u.

(2) *Choose a statistical test.* Theory associated with Student's *t* allows derivation of an expression linking the *test statistic t*

with the *parameter M*:

$$t = (M - \overline{M}) \times \sqrt{N}/\bar{s} \text{ with } \nu = N - 1 \quad (5.6)$$

where $\overline{M}$ is the researcher's *estimate* of the population mean, M is the true (and unknown) value of the population mean and $\bar{s}$ is an *estimate* of the population standard deviation.

As the probability density function of t is known, then so is the pdf of $(M - \overline{M})$ via Equation 5.6. Thus the probability of occurrence of various 'true' values of M may be calculated. In particular, consider the upper and lower bounding values M_u and M_l:

$$t_u = (M_u - \overline{M}) \times \sqrt{N}/\bar{s}$$

$$t_l = (M_l - \overline{M}) \times \sqrt{N}/\bar{s}$$

from which

$$M_u = \frac{t_u \bar{s}}{\sqrt{N}} + \overline{M} \quad (5.7a)$$

$$M_l = \frac{t_l \bar{s}}{\sqrt{N}} + \overline{M} \quad (5.7b)$$

(3) *Choose a sample size and a size for the critical region.* The sample size chosen is $N = 134$. Let the size of the critical region be $a = 0.01$. Now this figure $a = 0.01$ is also the probability of making a type 1 error (Section 2.4), i.e. the probability of rejecting the null hypothesis when, in fact, it is true. So conversely, the probability of accepting the null hypothesis when it is true is $(1 - a)$, i.e. 0.99. So $a = 0.01$ for a 0.99 confidence interval.
(4) *Investigate the sampling distribution of the test statistic.* This has already been described above. It is symmetrical about $t = 0$ and larger values are decreasingly likely.
(5) *Define the critical region.* In this case, we require our critical region to be split equally between the two tails of the t distribution because, as Equations 5.7a and b show, t_u will be positive, t_l negative, but they will be of equal magnitude. Noting that the number of degrees of freedom is $\nu = N - 1 = 133$, the two-tail critical value of t for $a = 0.01$ is given by Table 5.5 as about 2.60 so that $t_u = +2.60$ and $t_l = -2.60$.

(6) *The calculation and the decision.* Substituting these values in Equations 5.7a and b gives $M_u = 1.77$ and $M_l = 1.47$. Thus we state that the true value of the mean of the maximum diameters of the population of pebbles from which our sample was drawn lies in the interval 1.47 to 1.77 inches with probability 0.99. Alternatively, we are '99% confident' that the 'true' mean is in the interval 1.47 to 1.77 inches.

5.7 Student's *t* and the difference between two means

Another application of Student's *t* distribution allows us to test whether or not two sample means could have been drawn from the same population. In other words, if $\overline{M}_1$ and $\overline{M}_2$ are the estimated means, is the difference $\overline{M}_1 - \overline{M}_2$ significantly different from zero? The necessary equations for this are:

$$t = (\overline{M}_1 - \overline{M}_2) \times \sqrt{[N_1 \times N_2/(N_1 + N_2)]}/s_0 \quad \text{with} \quad \nu = N_1 + N_2 - 2 \qquad (5.8)$$

where s_0 is given by:

$$s_0^2 = [(N_1 - 1) \times \bar{s}_1^2 + (N_2 - 1) \times \bar{s}_2^2]/\nu$$

Exercise. A sample of 134 pebbles has a mean value of maximum diameters of 1.62 in with standard deviation 0.68 in. For a second sample of 90, the corresponding figures are 1.69 in and 0.62 in. With a critical region of size 0.01, show that you cannot reject the null hypothesis that the two samples are drawn from the same population.

5.8 Parametric and non-parametric tests

In this last exercise, you have used Student's *t* to test for the difference between two means. The *test statistic* was calculated from *estimates* of the population parameters *M* and *s* and thus made assumptions about the underlying distribution of the variable; namely that it was Gaussian. In the case of the larger sample, this assumption was tested earlier in this chapter and was found to be justified. This is in distinct contrast to the Kolmogorov–Smirnov

Table 5.5 Critical values of Student's t. The critical region contains values of t greater than the critical value.

ν	One-tail $a = 0.05$	One-tail $a = 0.01$	Two-tail $a = 0.05$	Two-tail $a = 0.01$
1	6.31	31.82	12.71	63.66
2	2.92	6.96	4.30	9.92
3	2.35	4.54	3.18	5.84
4	2.13	3.75	2.78	4.60
5	2.02	3.36	2.57	4.03
6	1.94	3.14	2.45	3.71
7	1.89	3.00	2.36	3.50
8	1.86	2.90	2.31	3.36
9	1.83	2.82	2.26	3.25
10	1.81	2.76	2.23	3.17
12	1.78	2.68	2.18	3.05
15	1.75	2.60	2.13	2.95
20	1.72	2.53	2.09	2.85
24	1.71	2.49	2.06	2.80
30	1.70	2.46	2.04	2.75
40	1.68	2.42	2.02	2.70
60	1.67	2.39	2.00	2.66

two-sample test where the *test statistic* was calculated directly from the cumulative distribution functions of the variables, i.e. *not* via parameters. Tests belonging to the former category are termed **parametric** and those belonging to the latter category, **non-parametric**. The difference is important because the former, although generally the more powerful, have the disadvantages of making assumptions about the distribution of the variable and assumptions about the scale on which the variable is measured.

5.9 Estimating proportions by counting – more confidence intervals

Lastly, there is another common context in which knowledge of a confidence interval is likely to be useful, namely in the estimation of proportions by counting. In the practical situation, such an exercise consists of counting the number of individuals R that possess a particular attribute in a sample of total size N. The ratio $R:N$ is then the estimated proportion of the special individuals in the population at large. Such estimates of proportions by counting feature commonly in microscopic petrography but the method is easily adapted to a variety of situations elsewhere.

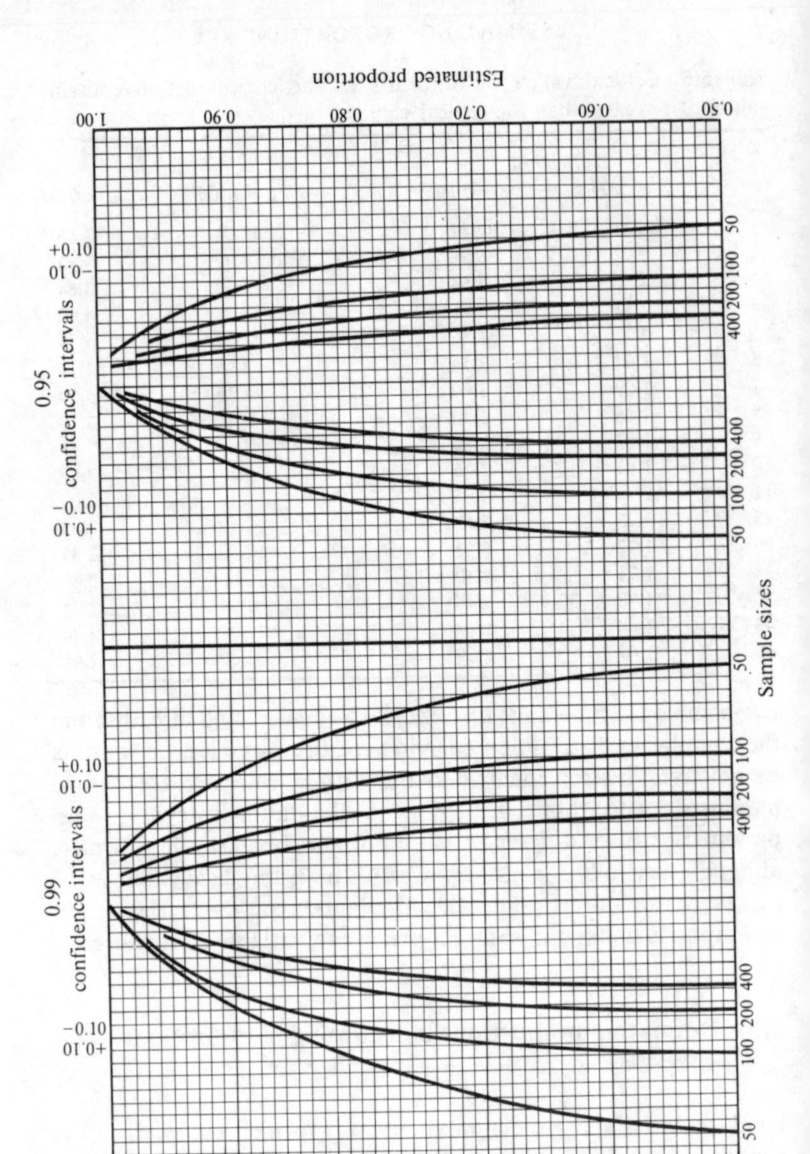

Figure 5.3 A chart for the calculation of confidence intervals of proportions estimated by counting. See text (Section 5.9) for a detailed example.

If the true proportion of special individuals in the population is H and in a sample of N individuals, we find a number R with the characteristic attribute, then our estimate of H is $R:N$. Now the probabilities of getting various values of R from a sample of N drawn from a population in which the true proportion is H can be calculated by Equation 2.1 (i.e. they are binomially distributed). Following six steps very similar to those outlined above (Section 5.6) for the calculation of confidence intervals surrounding an estimated mean, we can use Equation 2.1 to calculate values of R, the probabilities of occurrence of which are very small and which will thus fall into our critical regions. In the practical situation, such confidence intervals are rarely needed to a high order of precision, so we can use a graphical means of estimation, as shown in Figure 5.3.

The figure is divided into two parts, one for 0.95 confidence intervals, the other for 0.99 confidence intervals. In each part, two curves, an upper and a lower, are given for each of the sample sizes 50, 100, 200 and 400.

Here is an example to show how they are used. 100 fossils are found at a locality, 4 of which are echinoids. The estimated proportion of echinoids in the fossilised population is thus $4/100 = 0.040$. To quote a 0.99 confidence interval on this estimated proportion, find 0.040 on the horizontal axis and note the two points, the lower and the upper, at which a vertical through 0.040 intersects the two curves for samples of 100. Now read off the values on the left-hand vertical axis corresponding to these two points: they are -0.035 and $+0.050$. The required confidence interval thus ranges from $0.040 - 0.035$ to $0.040 + 0.050$, i.e. from 0.005 to 0.90 (loosely, I am 99% confident that the 'true' proportion of echinoids is somewhere between 0.5% and 9%). For estimated proportions over 0.50, note that the graphs need to be turned upside down. Note also that in trying to narrow the confidence interval by counting larger samples, there is a 'law of (rapidly) diminishing returns'. To halve the range of a confidence interval, it is necessary to quadruple the sample size.

6 Ordinal-scale methods

This chapter concerns itself with some of the statistical methods that may be introduced when the data consist of specimens that have simply been ranked in order of increasing value of some variable, i.e. measured on an ordinal scale. In such an ordering, the median value of the variable is easily found: it is the value associated with the specimen that occupies the midpoint of the ranking (or the average of those flanking the midpoint for samples of even size). Because of this consequent ease of locating the median, I include tests for difference of medians between two samples. Also introduced are minor variations that arise in those studies where a sample is split into two or more subsamples by virtue of specimens being measured on a nominal scale subsidiary to the ordinal scale, e.g. sedimentary beds of different thicknesses classified by rock type. Although the tests given here are all non-parametric, they deserve close attention because some approach the power of the better known parametric tests but without the disadvantages of some of the restrictive constraints attaching to the latter. Usefully, most need no more space than the back of the proverbial envelope for the computational work.

6.1 Classifying or measuring on two or more scales simultaneously

So far, we have thought of two-sample testing as involving the testing of one sample against another quite separate sample under some null hypothesis of equality. But as a minor variation, we could consider splitting a single sample into two or more subsamples by additionally measuring each specimen on some nominal scale and then testing these subsamples against each other. A simple example here is in the classification of pebbles according to rock type in a study of their lengths. Here there is a limited comparison with our discussion of porphyroblastic gneisses in Chapter 4 where each specimen (locality) was measured on two nominal scales (one, hornblende present or absent; the other, biotite present or absent). Such situations are common in practice, but to introduce the methods, I shall confine our attention to two scales of measurement

Table 6.1 Relations of various nominal and ordinal methods.

		Variable 2	
		Nominal	*Ordinal*
Variable 1	*Nominal*	contingency table (chi-squared test, Fisher's exact probability test)	Kolmogorov–Smirnov two-sample test
	Ordinal	median test, Wald–Wolfowitz runs test Mann–Whitney U test, Kruskal–Wallis one-way analysis of variance	Kendall's rank correlation coefficient τ

on each specimen, and our choice of scales to nominal and ordinal. So doing, Table 6.1 can be constructed to show the relative positions of some procedures that are useful. Note that all are non-parametric (see Section 5.8).

6.2 An example – possible differences between subsamples

To introduce the tests in the lower left of Table 6.1, and to help with their comparison, consider a sample of 72 beach pebbles that have been classified (i.e. measured on a nominal scale) into two subsamples: 48 of 'dolerite' and 24 'sedimentary'. One of several variables that is easily measured on an ordinal scale is 'length': the pebbles are arranged in order of increasing length, the shortest at one end labelled '1' and the longest at the other end labelled '72'. The outcome of this ordering is shown in Table 6.2.

With this ranking, it is easy to locate the median values: the median length of the pebbles of sedimentary rocks is 39 mm (the average length of pebbles 26 and 27), of the pebbles of dolerite 44 mm (the average length of pebbles 40 and 41), and of the total sample, 43 mm. These figures immediately suggest differences in length between 'sedimentary' and 'dolerite' pebbles. However, there are other ways in which the two subsamples may differ, as suggested by the hypothetical frequency distribution curves sketched in Figures 6.1a–d.

Exercise. Sketch the cdf's corresponding to the pdf's of Figures 6.1a–d.

The null hypothesis (of no differences) will be much the same in each case but alternative hypotheses may be framed in many ways according to the kinds of differences we may be interested in. It could be that the alternative hypothesis will refer to differences between the subsamples in only one of the parameters of Figure 6.1, or with respect to two of them, or with respect to all of them. The less specific is the alternative hypothesis then, generally, the less powerful is the appropriate statistical test. The tests

Table 6.2 Ordinal measure of 'length' in a sample of 72 beach pebbles, 48 of 'dolerite' (D) and 24 of 'sedimentary rocks' (S). 'Runs' (i.e. unbroken sequences) of 'D' pebbles and 'S' pebbles are separated by cross bars: $r = 30$; $R_{dol} = 1926$; $R_{sed} = 702$ (see text for explanation). The median length of 'S' pebbles is the average of that of numbers 26 and 27 (39 mm), the median length of 'D' pebbles is the average of that of numbers 40 and 41 (44 mm). The overall median length is the average of that of numbers 36 and 37 (43 mm).

Rank	*Rock type*	*Rank*	*Rock type*	*Rank*	*Rock type*
1	S	25	S	49	S
2	S	26	S	50	D
3	D	27	S	51	D
4	S	28	D	52	S
5	D	29	D	53	S
6	D	30	S	54	D
7	D	31	D	55	S
8	D	32	D	56	D
9	S	33	D	57	S
10	S	34	S	58	D
11	D	35	D	59	D
12	D	36	D	60	D
13	S	37	D	61	S
14	S	38	D	62	D
15	S	39	S	63	D
16	D	40	D	64	D
17	S	41	D	65	D
18	S	42	D	66	D
19	D	43	S	67	D
20	D	44	D	68	D
21	D	45	D	69	D
22	D	46	D	70	D
23	D	47	D	71	D
24	D	48	S	72	D

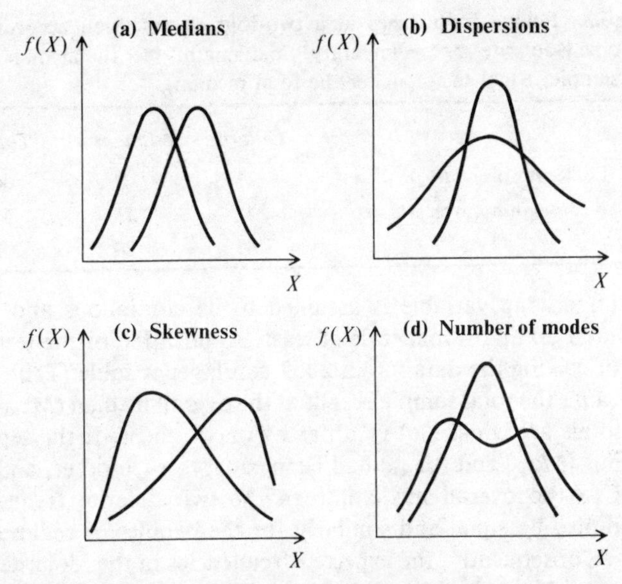

Figure 6.1 Probability density functions, illustrating some of the ways in which two populations may differ. All the curves are sketched on the same axes, the variable X along the horizontal axis, and its pdf $f(X)$ along the vertical axis. The sketches show: (a) identical curves, differing only in medians; (b) symmetrical curves with identical medians but different dispersions; (c) asymmetrical curves leaning left or right with similar dispersions but different medians; (d) symmetrical curves with similar medians and possibly similar dispersions but with different numbers of modes.

in the lower left of Table 6.1 are arranged approximately in order of increasing power and I shall introduce them accordingly. Assuming you are now familiar with the formalities of the six-step hypothesis-testing procedure, I shall describe each test only in the briefest outline. For all of these tests, the specimens are arranged in order of increasing magnitude of the variable, but retaining a labelling of the samples or subsamples similar to that used in Table 6.2.

6.3 The median test

For the median test, an appropriate form of the alternative hypothesis is to the effect that the two samples or the two subsamples differ with respect to their median values, with no reference made to any other of the population parameters and no reference made to the 'direction' of difference (see Section 2.5).

Table 6.3 Pebble frequencies in a two-fold classification according to rock type ('dolerite' or 'sedimentary') and length (over the median of the total sample; equal to, or under the total median).

	Dolerite	*Sedimentary*	*Totals*
'long' (i.e. length over Me_{tot})	27	9	36
'short' (i.e. length not over Me_{tot})	21	15	36
totals	48	24	72

The underlying variable is assumed to be continuous and to be measured on an ordinal scale at least. Application of the test consists of casting the data into a 2×2 contingency table (Table 6.3).

Because the total sample is split at the overall median (Me_{tot}), the row totals are equal (36) or differ by one at most. If the separate medians (Me_{dol} and Me_{sed}) had been equal to each other, and thus equal to the overall median, then the two dolerite frequencies would also be equal and similarly for the pebbles of sedimentary rocks. Consequently, the expected frequencies in the 'dolerite' column would both be 24 and in the 'sedimentary' column, both 12. However, the pairs of observed frequencies in the two columns are not equal and the question arises as to whether this inequality is a chance discrepancy under the null hypothesis true or whether it represents a significant discrepancy warranting rejection of the null hypothesis. As this is a 2×2 contingency table, it may be tested by the methods of Chapter 4. Because the sample size is over 50 and no expected frequency is under 10, the chi-squared test is applicable and, in this example, yields a value, $\chi^2 = 2.25$. With one degree of freedom (ν) and a critical region of size 0.05, the null hypothesis (of equal subsample medians) is accepted because the critical region contains values of χ^2 of 3.84 and greater.

The test splits the samples at their joint median, but makes no other use of the specimen values, so many of the data are made redundant and consequently the test lacks power. Also, because no reference can be made to the 'direction' of the difference between medians, no one-tail version exists.

6.4 The Wald–Wolfowitz runs test

When the two samples, or the two subsamples, are ranked together, as in Table 6.2, and the identities of the specimens are retained (e.g. whether the pebble is 'S' or 'D') then a number of unbroken sequences, or 'runs' arises as shown by the zigzag line. The length of a run may be as short as one or, at the other extreme, the two samples may be completely segregated, giving two runs only. More generally, there will be some intermixing which will reflect one or other of the frequency distributions of Figure 6.1. Thus a difference in dispersions about a common median will give long runs at the extremes of the overall ranking and a good deal of mixing in the centre. In Table 6.2, individual runs are separated by the cross-bars of the zigzag and there is a total of $r = 30$ runs. Apart from a longish run of 11 'D' pebbles at the top end of the ranking (pebble numbers 62 to 72), there seems fairly uniform mixing.

As its null hypotheses, the Wald–Wolfowitz runs test (or simply 'the runs test') takes the assumption that the two samples (or the two subsamples) do not differ in their distributions in any way, with respect to medians, dispersions, skewness, etc. Consequently, the test seeks any kind of difference between the two distributions and, because of this generality, is again likely to be a less powerful test. Because of this enforced generality of the alternative hypothesis, no one-tail version of the test can be constructed.

Mechanically, the theory behind the test predicts the number of unbroken runs that are expected when two samples or subsamples from identical populations are ranked together, as in Table 6.2. If the observed number of runs departs significantly from that predicted, then the null hypothesis is rejected. The test statistic is z, calculated according to Equation 6.1. Critical values of z appear in Table 5.3 (remember it is the two-tail values that are appropriate) and the critical region contains the larger values of z. The approximation given by Equation 6.1 suffices for all but ridiculously small sample sizes (nine or less!). For the current example, with 30 runs, $z = -0.80$, so the null hypothesis is not rejected for a critical region as large as 0.05.

If r is the observed number of runs, then the test statistic z is calculated, approximately, as follows:

$$\text{Let } N = N_1 + N_2;\ P = 2 \times N_1 \times N_2$$

where N_1 and N_2 are the sizes of the two (sub) samples.

$$\text{Then if: } M = 1 + (P/N)$$

$$\text{and: } s = \sqrt{\{P \times (P - N)/[N^2 \times (N - 1)]\}}$$

$$z = (r - M)/s \qquad (6.1)$$

6.5 The Mann–Whitney *U* test

The Mann–Whitney *U* test is a more powerful test, sensitive to differences of median values, and functions as follows. If two samples are drawn from the same population and one specimen is drawn at random from each sample, then the probability of the first specimen being larger than the second specimen is exactly 0.5. If every specimen in one sample is compared with every specimen in the other sample and the probabilities arising from these comparisons do not 'average out' to approximately 0.5, then it is likely that the two samples are drawn from populations that differ with respect to their medians. The power of this test is due to the fact that every specimen in one sample is effectively compared with every specimen in the other sample, thus making full use of the ordinal data. The test statistic *U* is calculated thus:

(1) The two samples or subsamples are combined and ranked according to increasing values of the variable, as in Table 6.2.
(2) The rank '1' is assigned to the smallest specimen, '2' to the next smallest, and so on, irrespective of to which sample the specimens belong.
(3) The ranks assigned to the first sample are totalled (R_1) and the ranks assigned to the second sample are totalled (R_2). In the example, $R_{\mathrm{dol}} = 1926$, $R_{\mathrm{sed}} = 702$.
(4) The quantities U_1 and U_2 are then calculated, but check first the note on evaluation of expressions at the beginning of this book:

$$U_1 = N_1 \times N_2 + N_1 \times (N_1 + 1)/2 - R_1$$

$$U_2 = N_1 \times N_2 + N_2 \times (N_2 + 1)/2 - R_2$$

(as a check: $U_1 + U_2 = N_1 \times N_2$)
In the example, $U_{\mathrm{dol}} = 402$, $U_{\mathrm{sed}} = 750$.
(5) The test statistic *U* is the smaller of U_1 and U_2 and for sample sizes over about 13 for the 0.01 critical region or over about 20 for the 0.05 critical region, *U* has an approximately Gaussian distribution with:

$$M = N_1 \times N_2/2$$

$$s = \sqrt{[N_1 \times N_2 \times (N + 1)/12]}$$

so that the test statistic $z = (U - M)/s$ may be calculated and critical values found in Table 5.3. The critical region will contain the values of z of greater magnitude (irrespective of sign) than the critical value. Alternatively, for sample sizes smaller than these, critical values of U may be found in Table 6.5.

In the example, $z = -2.08$, thus falling into a critical region of size 0.05. Consequently, the use of the more powerful Mann–Whitney U test leads to acceptance of the alternative hypothesis that the median length of the pebbles of dolerite is greater than the median length of the pebbles of sedimentary rocks.

6.6 Kruskal–Wallis one-way analysis of variance

The last and most powerful of these tests is the Kruskal–Wallis 'one-way analysis of variance'. We have met the term variance earlier (Section 5.2.1) but the context here is somewhat different. Previously, variance was used specifically for the square of the standard deviation in a Gaussian population and an expression was introduced by which it could be estimated. Here, a rather more general, non-quantitative meaning is being attached in that the Kruskal–Wallis test is sensitive to differences in medians and, to a lesser extent, differences in dispersion. Consequently, our variance here is not a rigidly defined quantity, it is a term used to introduce a wider concept of between-sample variation.

Analysis of variance is used in an attempt to discover the sources of variability between samples, or subsamples. In our current example, we have measured lengths of pebbles on an ordinal scale and have allocated the pebbles to one or the other of two subsamples on the basis of rock type. Is there variability between these subsamples with respect to length? The Kruskal–Wallis test is a non-parametric method of attempting to answer this question. If the answer is 'yes', then a source of variability should be sought. If 'no', then the samples, or subsamples, are said to be **homogeneous**.

'One-way' analysis of variance simply refers to the number of variables that are being considered during the operation. In the present example, there is only one variable, 'length'. Parametric methods of analysis of variance can use 'two-way', 'three-way', etc. mechanisms so that variability arising from two or more variables can be considered. However, these methods are rarely sufficiently convenient to use on the spot, so I refer interested readers to Sprent (1977) and Yule and Kendall (1950) for readable accounts.

The Kruskal–Wallis test is useful in that it can accommodate two *or more* samples (or subsamples). The procedure is very similar to that of the Mann–Whitney test, as far as step (3), the summing of the ranks for each of the samples or subsamples. Next, the test statistic H is calculated by Equation 6.2.

$$H = \frac{12}{N(N+1)} \times \sum_{i=1}^{i=S} (R_i^2/N_i) - 3 \times (N+1) \qquad (6.2)$$

where i labels the samples (or subsamples) from 1 to S, R_i is the sum of the ranks in the ith sample, N_i is the size of the ith sample and N is the combined sample size.

For N greater than about 50 and all N_i greater than about 10, H can be replaced by χ^2 with the number of degrees of freedom $\nu = S - 1$ so that critical values may be found in Table 4.6. In the example, $H = 4.32$. The critical region of size 0.05 for $\nu = S - 1 = 1$ contains all values of H of 3.84 and above, so that the null hypothesis is again rejected.

6.7 Comparison of parametric and non-parametric tests

To complete this introduction to combined nominal and ordinal tests, we note that the parametric Student's t test for differences between means, introduced in the last chapter, can be applied in the example above if the lengths of all specimens are measured on a ratio scale and assumed to have a Gaussian distribution. The resulting value, $t = 3.85$ warrants rejection of the null hypothesis for the much *smaller* two-tail critical region of size *0.001*. We have already noted that parametric tests are generally more powerful than equivalent non-parametric tests and the example can be taken to demonstrate this. Note, however, the much greater labour of calculation involved in the preliminary testing for Gaussian distributions. More important, note that the sample of pebbles was 'adjusted' to allow demonstration of the increasing power of the above tests. This 'adjustment' consisted of no more than removing the three largest 'sedimentary' pebbles from the original sample, previous to which, *none* of the tests (including Student's t) resulted in rejection of the null hypothesis even with a large 0.05 critical region. Although in this case the 'adjustment' was voluntary, *it could quite easily have been occasioned by a careless sampling procedure*. I hope I make clear in this way that for the more powerful statistical tests a much more carefully laid sampling plan is essential. More about this in Chapter 10.

6.8 Kendall's τ – a measure of ordinal-scale correlation

Finally in this chapter, I include a rapid method of comparing measurements on one ordinal scale with measurements on another ordinal scale with respect to their 'correlation'. To illustrate the meaning of **correlation**, suppose a sample of pebbles is arranged in a line with the longest at one end and the shortest at the other. Length is thus measured on an ordinal scale. Suppose this opera-

Table 6.4 Ordinal measurements of 'length' and 'width' in a sample of 24 dolerite pebbles.

Width ranks	*Length ranks*	*Larger*	*Smaller*	*Larger minus smaller*
1	2	22	1	+21
2	1	22	0	+22
3	3	21	0	+21
4	6	18	2	+16
5	7	17	2	+15
6	5	17	1	+16
7	17	7	10	− 3
8	12	11	5	+ 6
9	4	15	0	+15
10	14	9	5	+ 4
11	13	9	4	+ 5
12	8	12	0	+12
13	9	11	0	+11
14	11	9	1	+ 8
15	19	5	4	+ 1
16	18	5	3	+ 2
17	16	5	2	+ 3
18	15	5	1	+ 4
19	10	5	0	+ 5
20	21	3	1	+ 2
21	24	0	3	− 3
22	20	2	0	+ 2
23	23	0	1	− 1
24	22	0	0	0
				total $S = +184$

tion also results in the somewhat remote coincidence of all the pebbles also being arranged in order of increasing width. We should say that the correlation of length and width is perfect and in this case is 'positive' so that longer pebbles are also wider. If arranging the pebbles in order of increasing width completely reversed their order of increasing length, then again we should say that correlation of length and width is perfect, but this time is 'negative' so that longer pebbles are narrower. If rearranging the pebbles in order of increasing width totally randomised the length order, then length and width would be 'uncorrelated', the two variables having no apparent connection with each other.

A measure of correlation for ordinal scale measurements is Kendall's τ which ranges in value from $+1$ for perfect positive correlation through zero for uncorrelated variables to -1 for perfect negative correlation. To illustrate the means of calculation, an example is given in Table 6.4. The steps in the calculation are as follows:

(1) Arrange the specimens in increasing order of magnitude of the first variable (length) and label the objects with the resulting rank, 1 for the smallest up to N for the largest.
(2) Rearrange the specimens in order of increasing magnitude of the second variable (width) and record the rearranged order of the variable 1 (length) ranks as in the second column of Table 6.4.
(3) For each specimen (pebble) in turn, scan *down* column 2, starting at the chosen specimen, counting the number of ranks that are *larger*. Enter the results in column 3.
(4) Repeat step (3), this time counting the number of ranks that are *smaller*. Enter the results in column 4.
(5) Subtract 'smaller' from 'larger' and enter the results in column 5. Total the entries in column 5 and let this total be S.
(6) Kendall's τ is given by:

$$\tau = (2S)/[N \times (N-1)] \tag{6.3}$$

In the example, $S = 184$ and $N = 24$ so that $\tau = +0.67$. Thus 'length' and 'width' *appear* to be positively correlated in this sample. However, the possibility exists that the observed value of τ arose by sampling fluctuation in a population in which, in truth, 'length' and 'width' are uncorrelated (i.e. $\tau = 0$). Thus we need to test whether or not τ differs significantly from zero. For samples of size 10 or greater, we can use a test statistic z given by:

$$z = \tau \times \sqrt{\{[9 \times N \times (N-1)]/[2 \times (2N+5)]\}} \qquad (6.4)$$

(Check 'Evaluation of expressions' at the beginning of this book.) Critical values of z are given in Table 5.3, the critical region containing values of z larger than the critical value. In the example, $z = +4.59$ so that a null hypothesis involving $\tau = 0$ is rejected with a one-tail critical region of size 0.01.

6.9 'Ties' in ordinal measurement

In the application of all of the ordinal scale methods mentioned above, it is always possible that two or more specimens will 'tie' when being measured on an ordinal scale. Such ties are rarely troublesome in practice as they are so infrequent and when they occur in very small numbers, they may safely be ignored. In the unlikely event that they are abundant, the reader is referred to Siegel (1956) for appropriate correction procedures.

Table 6.5 Critical values of U in the Mann–Whitney test.

N_2 \ N_1	3		4		5		6		7		8		9		10	
3	0	–														
	–	–														
4	0	–	1	0												
	–	–	–	–												
5	1	0	2	1	4	2										
	–	–	0	–	1	0										
6	2	1	3	2	5	3	7	5								
	–	–	1	0	2	1	3	2								
7	2	1	4	3	6	5	8	6	11	8						
	0	–	1	0	3	2	4	3	6	4						
8	3	2	5	4	8	6	10	8	13	10	15	13				
	0	–	2	1	4	3	6	4	8	6	10	8				
9	3	2	6	4	9	7	12	10	15	12	18	15	21	17		
	1	0	3	1	5	3	7	5	9	7	11	9	14	11		
10	4	3	7	5	11	8	14	11	17	14	20	17	24	20	27	23
	1	0	3	2	6	4	8	6	11	9	13	11	16	13	19	16
11	5	3	8	6	12	9	16	13	19	16	23	19	27	23	31	26
	1	0	4	2	7	5	9	7	12	10	15	13	18	16	22	18
12	5	4	9	7	13	11	17	14	21	18	26	22	30	26	34	29
	2	1	5	3	8	6	11	9	14	12	17	15	21	18	24	21
13	6	4	10	8	15	12	19	16	24	20	28	24	33	28	37	33
	2	1	5	3	9	7	12	10	16	13	20	17	23	20	27	24
14	7	5	11	9	16	13	21	17	26	22	31	26	36	31	41	36
	2	1	6	4	10	7	13	11	17	15	22	18	26	22	30	26
15	7	5	12	10	18	14	23	19	28	24	33	29	39	34	44	39
	3	2	7	5	11	8	15	12	19	16	24	20	28	24	33	29
16	8	6	14	11	19	15	25	21	30	26	36	31	42	37	48	42
	3	2	7	5	12	9	16	13	21	18	26	22	31	27	36	31
17	9	6	15	11	20	17	26	22	33	28	39	34	45	39	51	45
	4	2	8	6	13	10	18	15	23	19	28	24	33	29	38	34
18	9	7	16	12	22	18	28	24	35	30	41	36	48	42	55	48
	4	2	9	6	14	11	19	16	24	21	30	26	36	31	41	37
19	10	7	17	13	23	19	30	25	37	32	44	38	51	45	58	52
	4	3	9	7	15	12	20	17	26	22	32	28	38	33	44	39

The table is divided into a series of boxes. Each column of boxes corresponds to a given number of specimens in sample 1 (N_1) each row of boxes to a given number of specimens in sample 2 (N_2). The test should be arranged so that N_1 is smaller than N_2 if the two are not equal. Within each box, there are four critical values of U, arranged thus:

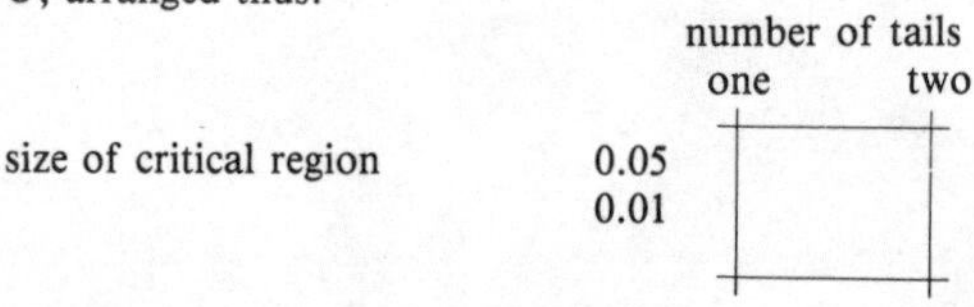

11	*12*	*13*	*14*	*15*	*16*	*17*	*18*	*19*
34 30 25 21								
38 33 28 24	42 37 31 27							
42 37 31 27	47 41 35 31	51 45 39 34						
46 40 34 30	51 45 38 34	56 50 43 38	61 55 47 42					
50 44 37 33	55 49 42 37	61 54 47 42	66 59 51 46	72 64 56 51				
54 47 41 36	60 53 46 41	65 59 51 45	71 64 56 50	77 70 61 55	83 75 66 60			
57 51 44 39	64 57 49 44	70 63 55 49	77 67 60 54	83 75 66 60	89 81 71 65	96 87 77 70		
61 55 47 42	68 61 53 47	75 67 59 53	82 74 65 58	88 80 70 64	95 86 76 70	102 93 82 75	109 99 88 81	
65 58 50 45	72 65 56 51	80 72 63 57	87 78 69 63	94 85 75 69	101 92 82 74	109 99 88 81	116 106 94 87	123 113 101 93

The critical region contains values of U equal to, or *smaller* than, the critical value.

7 Correlation and regression

It might be unusual for the methods I mention in this chapter to find application 'on site', as is the theme of this book. The twofold reasons for this are the labour of ratio-scale measurement and the labour of calculation involved. My primary purpose for their brief introduction lies in the link they provide with sophisticated, generally computer-based methods, and to a lesser extent, their potential application in elementary laboratory studies.

7.1 Univariate, bivariate and multivariate populations

Each specimen in a sample may be measured on more than one scale. We have already seen an example in the measurement of a pebble according to its length and its rock type; one measurement on a ratio scale, the other on a nominal scale. In this case, because we measure two variables, we are said to be sampling a 'bivariate' population. Most examples in the book relate to 'univariate' (one-variable) populations. Strictly speaking, we are not restricted by the type of scale of measurement (nominal, ordinal or ratio), but complications may arise if we do mix scales. Consequently, common applications of bivariate and multivariate correlation and regression methods are restricted to ratio-scale measurements. Examples of such a study might be a comparison of various dimensions (e.g. length, width, height, etc.) of a sample of fossils thought to belong to a single species, or of the distribution of chemical components in a suite of rocks. The latter study, however, might be complicated by the fact that the sum of chemical components for any given rock is constant (usually taken as 100%). Such problems are discussed by Chayes (1971) and Davis (1973).

7.2 Correlation

A scatter diagram (e.g. Fig. 7.1) is a convenient means of graphical representation of a bivariate sample or two of the variables from a multivariate sample. Each specimen is represented as a point, the

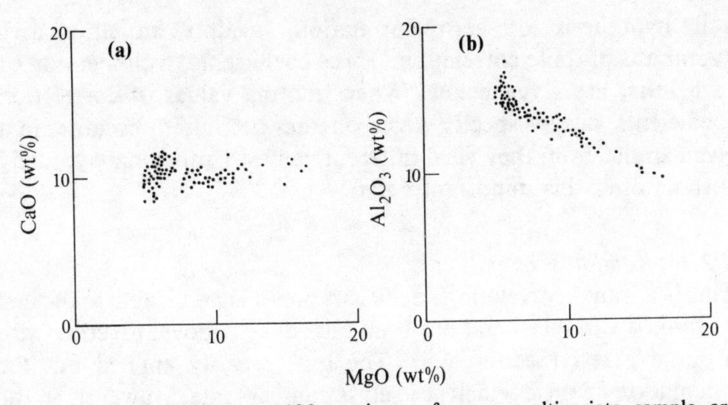

Figure 7.1 Scatter diagrams. 106 specimens from a multivariate sample are represented by points whose horizontal co-ordinate is proportional to one variable and whose vertical co-ordinate is proportional to a second variable. The sample is of chemical analyses of basaltic lavas from Mauritius and the data are kindly provided by Dr A. N. Baxter. In (a) Pearsons's r has the value -0.09 and is not significantly different from zero. CaO and MgO concentrations are thus uncorrelated. In (b) the same coefficient has the value -0.93, indicating good negative correlation between Al_2O_3 and MgO concentrations. Note that specimens with under 5% by weight of MgO have been omitted on petrological grounds.

horizontal and vertical co-ordinates of which are proportional to the two ratio-scale measurements on the specimen. The disposition of the array of points on such a diagram may immediately suggest whether there is any connection (or 'correlation') between the values of the variables displayed. Thus in Figure 7.1b, we see that, on average, the Al_2O_3 content of rocks in this particular sample decreases as the MgO content increases. This is 'negative correlation'. Were the two variables to increase jointly, the case would be one of 'positive correlation'. In Figure 7.1a, there seems, on average, neither increase nor decrease of CaO content as MgO content rises. Here, we may suspect the variables are 'uncorrelated'.

It is useful to be able to give an indication of the strength and the sign (positive or negative) of correlation between two variables. Ideally, such a 'coefficient of correlation' should range from a maximum of $+1$ (for perfect positive correlation) through zero (for uncorrelated variables) to a minimum of -1 (for perfect negative correlation). Further, it should be an elementary matter to test whether or not any observed value of a correlation coefficient differs significantly from zero in case a spurious finite value arises in a sample from a population in which, in fact, the variables are uncorrelated. In short, to test whether or not one may reject the null hypothesis of zero correlation, against an alternative hypothesis of finite correlation. Three coefficients, with mention of such tests, are given below. When quoting values of correlation coefficients, always specify which one has been used, because, in a given application, they yield different values. Unfortunately, many authors omit this important detail.

7.2.1 Kendall's τ

This is a 'rank' correlation coefficient (i.e. it is applicable to ordinal or ranked variables) and has been introduced above, together with a suitable test (Section 6.8). The test is easily applied but the calculation of the coefficient itself is cumbersome. However, in the case of multivariate samples, it has an advantage. In such samples, it is always possible that variables x and y show a strong correlation not because they are intrinsically linked, but because they both show separate strong correlations with a third variable z. Observers should be wary of this contingency which can be approached by use of Kendall's 'partial' rank correlation coefficient, calculated by methods akin to those mentioned above. Details are outside the scope of this book, but refer to Yule and Kendall (1950, pp.258–64 and p.306) and to Siegel (1956, pp.223–9).

7.2.2 Spearman's rank correlation coefficient

Like Kendall's τ, this is a coefficient which has the advantage of easier calculation, but the double disadvantage of no available test and no equivalent partial form. See Yule and Kendall (1950) for details.

7.2.3 Pearson's r

This coefficient is applicable to ratio-scale measurements, and theorists find that it is readily manipulated algebraically, so it has received much attention. However, in addition to the labour of ratio-scale measurement the calculation is protracted, although the method is outlined in Table 7.1. Whether or not the observed value of the coefficient differs significantly from zero may be tested by calculation of Student's t:

$$t = r \times \sqrt{[(N-2)/(1-r^2)]} \tag{7.1}$$

Critical values of t are found in Table 5.5, the number of degrees of freedom, $\nu = N - 2$.

Table 7.1 Calculation of Pearson's correlation coefficient r and of regression constants and fiducial limits.

Let the regression line have the equation $y = a + bx$. The calculation is based on a sample $N = 30$ of Al_2O_3/MgO analyses used in the construction of Figure 7.1b. Let the MgO concentration be the independent variable x and the Al_2O_3 concentration be the dependent variable y.

(1) *Calculate the following quantities*		$N = 30$
sum of x's	$\Sigma(x)$	217.74
sum of y's	$\Sigma(y)$	427.36
sum of squares of x's	$\Sigma(x^2)$	1780.5104
sum of squares of y's	$\Sigma(y^2)$	6151.3084
sum of products xy	$\Sigma(xy)$	3000.9713
(2) *From these, derive:*		
mean of x's	$\overline{M}_x = \Sigma(x)/N$	7.258
mean of y's	$\overline{M}_y = \Sigma(y)/N$	14.245
variance of x	$s_x^2 = \Sigma(x^2)/N - \overline{M}_x^2$	6.67
variance of y	$s_y^2 = \Sigma(y^2)/N - \overline{M}_y^2$	2.19
covariance of x and y	$s_{xy} = \Sigma(xy)/N - \overline{M}_x\overline{M}_y$	-3.36
correlation coefficient r	$r^2 = 1 - \left(\dfrac{N-1}{N-2}\right)\left\{1 - \dfrac{s_{xy}^2}{s_x^2 s_y^2}\right\}$	0.79
	r(sign from covariance)	-0.89
residual variance S_y^2	$S_y^2 = s_y^2(1 - r^2)$	0.67
regression intercept a	$a = \overline{M}_y - \overline{M}_x s_{xy}/s_x^2$	17.9
standard error e_a	$e^2 = S_y^2/N$	± 0.12
regression slope b	$b = s_{xy}/s_x^2$	-0.50
standard error e_b	$e^2 = S_y^2/(s_x^2 N)$	± 0.047

(3) *For convenient values of* x, *calculate corresponding values of* S_r:

$$S_r^2 = S_y^2[1 + (x - \overline{M}_x)^2/s_y^2]/N$$

These give upper and lower fiducial limits to predicted y of:

$$a + bx \pm S_r t$$

where Student's t comes from Table 5.5 with $v = N - 2$ degrees of freedom.

7.3 Regression

Patterns of points on scatter diagrams, or calculations of correlation coefficients, or both, may suggest that there is some 'functional' relationship between the two variables, i.e. a relationship that may be expressed in the form of an equation. Usually this will be a 'linear' equation, i.e. an equation whose graph is a straight line:

$$y = a + (b \times x) \tag{7.2}$$

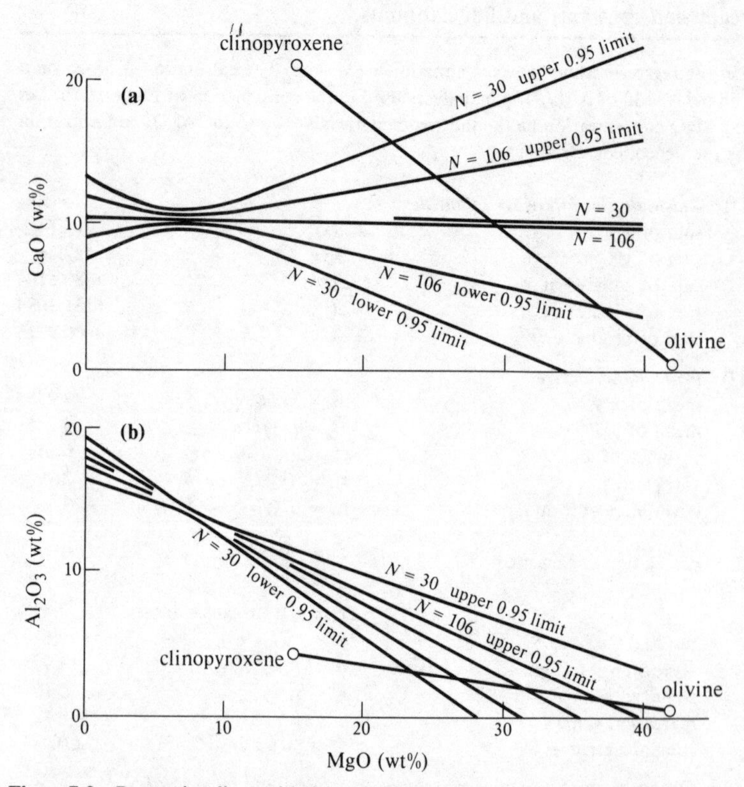

Figure 7.2 Regression lines with fiducial limits, based on the same data as Figure 7.1. These basaltic lavas contain phenocrysts of olivine and clinopyroxene, the positions of which, in terms of their CaO, Al_2O_3 and MgO concentrations, are shown. From the dispersions of points displayed in Figure 7.1, it may be suggested the lavas have evolved by the differentiation of a certain but unknown proportion of olivine and pyroxene. Assuming a linear (i.e. straight line) process and taking MgO concentration as an index of differentiation, a regression line extrapolated to the olivine–clinopyroxene join will give an estimate of the bulk composition of the precipitating phases. Regression lines have been calculated for the full sample of 106 specimens and also for a subsample, randomly chosen, of 30 specimens. Computation of the 0.95 (i.e. 95%) fiducial limits has been included in the calculation for purposes of comparison. Notice how the fiducial band (the distance between upper and lower fiducial limits) increases dramatically in width with extremes of extrapolation. The intersections of the regression lines and their fiducial bands with the olivine–clinopyroxene joins allows estimation of relative proportions of precipitating crystals.

in which *a* and *b* are constants called the 'regression intercept' (i.e. the intercept value of the straight line on the vertical axis) and the 'regression slope (or coefficient)', respectively. *x* is said to be the 'independent' variable, and *y* the 'dependent' variable (dependent on *x*). If we suspect the relation is non-linear, then we can replace either or both of the variables *x* and *y* by some function (e.g. log (x), $\sqrt{y}$) so that Equation 7.2 becomes linear.

The task of calculating the values of the constants *a* and *b* constitutes the central part of a 'regression' analysis, so called because one of the first applications was to a statistical study of regressive tendencies between fathers and sons. The method is based on the 'least squares' approach, in which *a* and *b* are chosen so that the sum of the squares of the differences between observed and predicted values of the dependent variables is a minimum.

In symbols: $\sum [y_{\text{obs}} - (a + bx)]^2$ is a minimum.

The scheme of the calculation is set out in Table 7.1 and the results are shown graphically in Figure 7.2.

Having calculated the values of the regression constants, they may be used simply as descriptive parameters, in which case, their standard errors of estimation should be calculated and quoted too. Multiplication of these standard errors by the factors 1.96 or 2.58 (Table 5.3) will allow the user to construct 0.95 or 0.99 confidence intervals, respectively.

Alternatively, the regression constants can be substituted in Equation 7.2 together with an *x* value, to predict a *y* value. In this case, the user will wish to know the confidence interval surrounding the predicted *y* value. The calculation is cumbersome, but is shown in Table 7.1. The upper and lower confidence limits may be calculated for a wide variety of *x* values and the results may be plotted on a graph together with the regression line itself. This has been done in Figure 7.2 and the resulting curves constitute **fiducial limits** to predicted *y* values.

7.4 Applications and developments

The algebraical convenience of ratio-scale correlation has already been mentioned. Together with the development of digital computing and the mechanisation or semi-mechanisation of ratio-scale measurement techniques in many areas of application, the statistical processing of large multivariate samples has received much attention, especially in commercial ventures. As a consequence, the literature has multiplied and a large variety of multivariate methods has been employed. Although these are extremely important methods, most, for one reason or another, are inaccessible to the geologist 'on site', so are not treated here. A convenient introductory text is that of Hope (1968).

8 Two-dimensional orientations

We looked briefly at the nature of orientation data in Section 1.2.5, defined 'orientation', 'direction' and 'axis' and listed some examples. For orientation data in two dimensions (on the circle) the most convenient form of graphical presentation is usually a circular plot, introduced in Section 1.3.7. We also saw that certain linear data that are cyclical may be transformed into orientations in two dimensions and treated as angles. From this starting point, this chapter deals with various tests that are applicable to circular data but not until we have looked more closely at some of their special features, particularly the concept of 'uniformity'. All the tests can be framed in terms of the six steps of Section 2.3. Some are non-parametric, but in certain conditions parametric tests may be based on theoretical distributions with the bonus of increased power. A few of the methods lean on a knowledge of trigonometry, but not too heavily, and make a little use of elementary vector algebra. However, the effort needed to master these is surely well spent because over the span of geology as a whole, orientations comprise much of the data. Before evaluating any of the expressions in this chapter, check the note at the beginning of the book.

8.1 Classes of tests, uniform distribution and preferred orientation

As with linear scales of measurement, our tests are going to be both non-parametric and parametric, one-sample and two-sample, one-tail and two-tail. In addition, when estimating values of population parameters, we shall find it useful to be able to calculate confidence intervals. However, before introducing these, there is one new concept, that of the **uniform distribution**, which must be mentioned. In the context of orientation measurements, we may think of a situation where the density of directions is much the same over the total circumference of the circle (in the two-dimensional case) or over the total surface of the sphere (in the three-dimensional case) when we picture the directions as radiating from the centre of the circle or sphere. If our samples are not 'uniformly distributed' in this sense, then they must show one or more significant clusters or other indications of having 'preferred orientations'. Consequently, in addition to the classes of tests mentioned above, we recognise a new class: tests of uniformity. In all of these tests of uniformity, the null hypothesis states that the sample is drawn from a population that has a uniform density distribution, but the individual tests are sensitive to different kinds of departures from uniformity. I shall introduce first three non-parametric tests of uniformity, then the von Mises distribution function and then various parametric methods based on this distribution. A reference text for the majority of these is the excellent book by Mardia (1972), although they are presented in a mathematical form perhaps not immediately accessible to many geologists.

The application of nearly all the tests in this chapter benefits from the preparation of a circular plot, which is a logical first step in all cases (see Section 1.3.7).

8.2 Tests of uniformity

8.2.1 The Hodges–Ajne test

A rapidly applied, but not powerful, test of uniformity is that due to Hodges and Ajne. The method requires a circular plot on which we locate a diameter, choosing its direction such that a minimum number of points (m) falls to one side. In the case of the goniatites in Figure 1.10, this diameter is approximately west–east and has

Table 8.1 Critical values of m in the Hodges–Ajne test. The critical region contains values of m equal to, or *smaller* than the critical value.

m	$a = 0.05$	$a = 0.01$
0	$N =$ 0–11	$N =$ 0–14
1	12–14	15–17
2	15–17	18–21
3	18–20	22–24
4	21–23	25
5	24–25	≈ 30
7	≈ 30	≈ 35
9	≈ 35	≈ 40
10	≈ 40	≈ 45
12	≈ 45	≈ 50
14	≈ 50	

$m = 11$ points on one side, 25 on the other. m becomes the test statistic and critical values are given in Table 8.1. Note we have to state the alternative hypothesis carefully because the test is not sensitive to departures from uniform distributions towards those that have bilateral symmetry.

8.2.2 *Kuiper's test*

A more powerful test has been devised by Kuiper, and the data manipulation is illustrated in Table 8.2 and Figure 8.1. The working of the test is similar to that of the Kolmogorov–Smirnov one-sample test in the case of linear data. First, the data are arranged in order of increasing magnitude, 'modulo 360°' (i.e. if the data are not repeated every 360°, they should be suitably transformed: for axial data, see Section 1.3.9, the angles should be doubled). Preparation of a circular plot simplifies this first step. Data points are then numbered from $i = 1$ to $i = N$, as in columns 1 and 2 of Table 8.2. At this stage, there is likely to be a saving of time if a graph such as Figure 8.1 is prepared. The straight line is the graph of $U_i = x_i/360$ as a function of X_i where the X_i are the individual directions. The line embodies the theoretical prediction

Table 8.2 Orientation of the apertures of 36 goniatites in a rock bed. A circular plot of the data appears in Figure 1.5. i labels the data points in ascending order of magnitude of the angles X_i.

$X°$	i	U_i	i/N	$U_i - i/N$	$X°$	i	U_i	i/N	$U_i - i/N$
1	1				176	19			
8	2				188	20			
12	3				198	21			
14	4				214	22			
17	5				236	23			
21	6				245	24			
23	7				259	25			
32	8				274	26	0.761	0.722	+0.039
36	9				294	27	0.817	0.750	+0.067*
47	10	0.131	0.278	−0.147	304	28	0.844	0.778	+0.067
58	11	0.161	0.306	−0.144	313	29	0.869	0.806	+0.064
65	12	0.181	0.333	−0.153	320	30	0.889	0.833	+0.056
72	13	0.200	0.361	−0.161	330	31	0.917	0.861	+0.056
81	14	0.225	0.389	−0.164*	332	32			
98	15	0.272	0.417	−0.144	339	33			
119	16	0.331	0.444	−0.114	345	34			
146	17				349	35			
164	18				354	36			

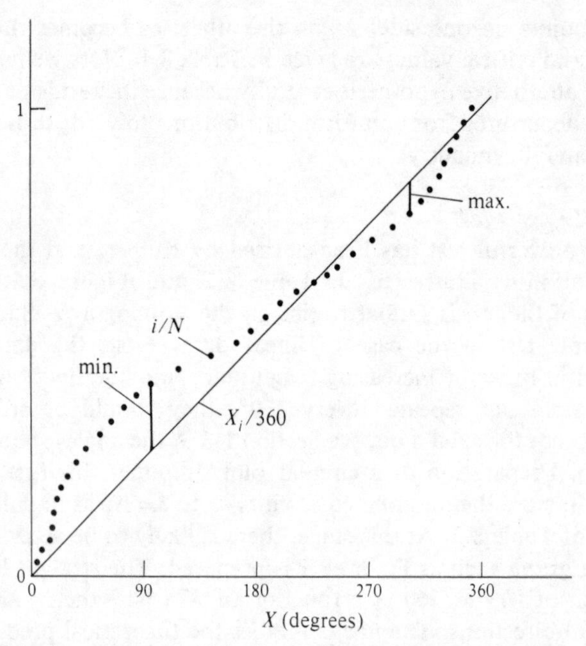

Figure 8.1 A sample arranged for the application of Kuiper's test. The horizontal axis measures X, transformed to run from zero to 360°. The vertical axis runs from zero to 1 and facilitates plotting if it is N units (e.g. mm) high. The straight line is drawn from the origin to the point with horizontal co-ordinate 360° and vertical co-ordinate 1. Points are i/N values for each specimen direction. Their horizontal co-ordinate is given by their associated direction X_i, their vertical co-ordinate by their i/N value or by their i value if the vertical axis is made N units high. min. is the largest of distances of points *above* the line, max. is the largest of distances *below* the line.

of uniform density distribution against which we are to test the sample. The array of points shows how the quantity i/N varies with values of X_i (the array being easily plotted if the vertical scale is made N units high). Clearly, if the sample were perfectly uniformly distributed, then i/N for any point would equal $X_i/360$ for the same point and the points would fall on the straight line. The more the distribution departs from uniformity, so the further are the points from the line. The test statistic is defined by:

$$V_N = \max(U_i - i/N) - \min(U_i - i/N) + 1/N \qquad (8.1)$$

where 'max' means the largest positive value and 'min' means the largest negative value.

A graph such as shown in Figure 8.1 can be used to derive

Table 8.3 Critical values of the test statistic V_N, the critical region containing observed values equal to or larger than the critical values.

$$\text{critical } V_N = V_N^*/[\sqrt{N} + 0.155 + 0.24/\sqrt{N}] \qquad (8.2)$$

critical region	$a = 0.05$	0.01
V_N^*	1.75	2.00

approximate values of V_N. If more-accurate values are necessary, then calculation has to be made as in Table 8.2, the graph indicating approximately which parts of the table need to be completed in order to find the 'max' and 'min' values for Equation 8.1, marked by asterisks in Table 8.2. In the example,

$$V_N = [+0.067 - (-0.164) + 1/36] = 0.259$$

Critical values of V_N are given in Table 8.3 (Stephens 1970). From this table, we find a critical value of 0.282 for a critical region of size 0.05, so decide to accept the null hypothesis of uniformity of distribution of the goniatite directions.

8.2.3 Watson's U^2 test

A non-parametric test, seemingly of even greater power, is Watson's U^2 test, but with the disadvantage of the labour of calculation entailed. As above, data are arranged in ascending order of magnitude, modulo 360°, and the test statistic calculated directly from them as follows:

$$U^2 = \sum_{i=1}^{i=N}(U_i^2) - \frac{2}{N}\sum_{i=1}^{i=N}(i \times U_i) + \bar{U} + (\bar{U} - \bar{U}^2 + 1/12) \times N \qquad (8.3a)$$

where U_i is the same as for Kuiper's test (i.e. $X_i/360$) and $\bar{U}$ is the arithmetic mean of the U_i values:

$$\bar{U} = \frac{1}{N}\sum_{i=1}^{i=N}(U_i) \qquad (8.3b)$$

Again, this test statistic increases in magnitude as the sample departs further from uniformity. If we calculate the modified statistic:

$$U^{*2} = [U^2 + (1/N^2 - 1/N)/10] \times (1 + 0.8/N) \qquad (8.4)$$

then we have critical values as in Table 8.4 (Stephens 1970).

Table 8.4 Critical values of the modified test statistic U^{*2}, the critical region containing the larger values. The approximation in Equation 8.4 holds for N over about 8.

	$a = 0.05$	$a = 0.01$
$U^{*2} =$	0.187	0.267

In the example, the observed value of U^{*2} is 0.207, so that the hypothesis of uniformity may be rejected with a critical region of size 0.05.

8.2.4 Wald–Wolfowitz runs test

The list of reasonably rapid non-parametric tests may be completed by giving brief mention to the Wald–Wolfowitz two-sample runs test, mentioned in detail in Section 6.4. If the two samples are clearly differentiated on the circular plot, the number of runs can be determined rapidly, starting from a point at the coincidence of two runs.

8.3 The von Mises distribution

8.3.1 The pdf and parameters, estimation

Probably the most important theoretical distribution on the circle is that of von Mises, occupying a position analogous to that of the Gaussian distribution for linear data. Its probability density function is given here, for reference only, by:

$$\mathrm{f}(X) = \exp\left[k \times \cos(X - X_0)\right] / \left[2 \times \pi \times I_0(k)\right] \qquad (8.5)$$

where the *parameters* of the distribution are X_0, the **mean direction**, and k, the **concentration,** discussed below. X is the circular variable (modulo 360°) and $I_0(k)$ is a constant (a Bessel function about which we have no need to worry) dependent on k. exp (a) means exponential e (the base of natural logarithms) raised to the power (a).

For illustration, the pdf's of two von Mises distributions are shown on a polar diagram in Figure 8.2. They are bilaterally symmetrical about the mean direction and each has one mode (probability density a maximum) and one anti-mode (probability density a minimum). The concentration parameter k may vary from zero to plus infinity and as it increases, the von Mises pdf departs further from a circle. Consequently, k can be compared to

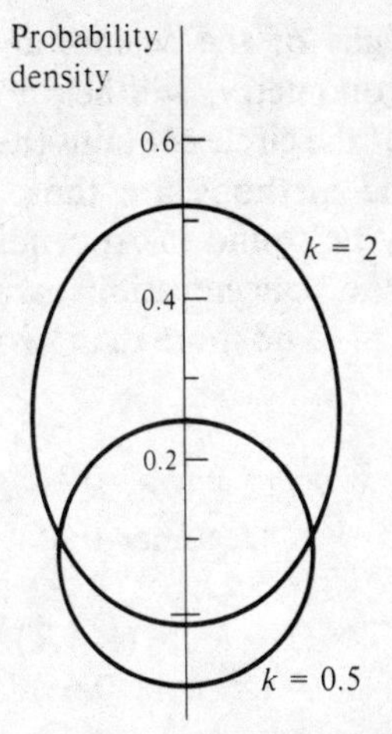

Figure 8.2 The pdf of the von Mises distribution. Two examples are shown on a polar diagram, one for concentration parameter $k = 0.5$, the other for $k = 2$. Note increase of eccentricity with increase in k.

the dispersion parameters of linear data (e.g. standard deviation) but also may be used as a measure of 'preferred orientation', the higher k the greater the preferred orientation. If a sample is assumed to be from a von Mises population, then the parameters may be estimated as follows:

(1) Find the mean values of the cosines and the sines of the sample directions thus:

$$\bar{C} = \frac{1}{N} \sum_{i=1}^{i=N} (\cos X_i) \tag{8.6a}$$

$$\bar{S} = \frac{1}{N} \sum_{i=1}^{i=N} (\sin X_i) \tag{8.6b}$$

(2) Calculate the 'mean resultant length':

$$\bar{R} = \sqrt{(\bar{C}^2 + \bar{S}^2)} \tag{8.7}$$

(3) Calculate the cosine and the sine of the estimate of the mean direction:

$$\cos \bar{X}_0 = \bar{C} / \bar{R} \tag{8.8a}$$

$$\sin X_0 = \bar{S} / \bar{R} \tag{8.8b}$$

Note that the signs of the cosines and sines, together with some simple trigonometry, will help to identify which of the four quadrants of the circle contains the estimated mean direction. So far, the methods are those of elementary vector algebra and may be found in Appendix A.

(4) For estimating the concentration parameter, crude approximations, sufficient to no more than two significant figures, are given by:

$$\bar{k} = \bar{R} \times (12 + 6 \times \bar{R}^2 + 5 \times \bar{R}^4)/6 \qquad (8.9a)$$
$$(\bar{R} \text{ under } 0.65)$$

$$1/\bar{k} = 2 \times (1 - \bar{R}) - (1 - \bar{R})^2 - (1 - \bar{R})^3 \qquad (8.9b)$$
$$(\bar{R} \text{ over } 0.65)$$

Fortunately, precise estimates are rarely needed.

For the goniatite directions of Table 8.2, the values of $\bar{C}$ and $\bar{S}$ are +0.313 and +0.007 respectively which, giving $\bar{R} = 0.313$, predict a mean direction of 0.5° measured clockwise from north, comparing favourably with the median direction of Figure 1.5. Equation 8.9a gives $\bar{k} = 0.66$.

Although many naturally occurring distributions on the circle are found to approximate well to a von Mises form, the statistical testing of this fit is involved because of the complicated calculation of the pdf or cdf. Consequently, we are likely to assume that we have a sample from a von Mises population if our circular plot is approximately bilaterally symmetrical and shows a reasonably smooth decrease in density from the mode to the anti-mode. An assumption of an underlying von Mises distribution seems realistic for the samples displayed in Figures 1.5 and 1.8. With the assumption of an underlying von Mises distribution, then a number of useful parametric methods becomes available, as described in the next section.

8.3.2 Tests based on the distribution

The first of these is a parametric test of uniformity due to Rayleigh. This test takes $\bar{R}$ as its test statistic and tests the null hypothesis of a uniform distribution (with $\bar{R} = 0$) against an alternative that the parent population is von Mises (with $\bar{R}$ significantly different from zero). Critical values of $\bar{R}$ are given in Table 8.5.

For the sample of 36 goniatites, the critical region of size 0.05

Table 8.5 Critical values of $\bar{R}$ in the Rayleigh test of uniformity. The critical region contains the larger values of $\bar{R}$ and the approximations are sufficient for N over about 15.

	$a = 0.05$	$a = 0.01$
critical R	$\sqrt{3.00/N}$	$\sqrt{4.61/N}$

contains values of $\bar{R}$ of 0.29 and above, so the null hypothesis is rejected since the observed value is 0.313.

Having estimated a mean direction, we are likely to want to specify a corresponding confidence interval after the style of Section 5.5. For moderate sample sizes with modest preferred orientation, calculation of a confidence interval for the mean direction is simple, the expression below being satisfactory for samples in which the product $N \times \bar{R} \times \bar{k}$ is not less than 6:

$$d° = K/\sqrt{(N \times \bar{R} \times \bar{k})} \qquad (8.10)$$

where the constant K is given by:

$1 - a$:	0.99	0.95
K:	148	112

The confidence interval then extends from $\bar{X}_0 - d°$ to $\bar{X}_0 + d°$.

In the goniatite example with $N = 36$, $\bar{R} = 0.31$ and $\bar{k} = 0.66$, $d = 41°$ for 0.95 confidence. Thus the confidence interval extends 41° to either side of the estimated mean direction.

In the two-sample situation, we are likely to want to test the null hypothesis that the two samples are drawn from the same von Mises population against the alternative that they are drawn from von Mises populations that differ either with respect to mean direction or with respect to concentrations. Both of these tests require that we calculate the mean resultant length of the *combined* samples. In this, we have to remember that the mean resultants of the two samples are *vectors*, so our arithmetic must treat them as such, by using the parallelogram rule of vector addition (Fig. 8.3). Using $\bar{C}$, $\bar{S}$ and $\bar{R}$ as above (Eqs 8.6a, b and 8.7) for the values applying to the *combined* sample, and appending the suffixes '1' and '2' for the corresponding quantities from the two separate samples, the rules of vector addition give:

$$N \times \bar{C} = (N_1 \times \bar{C}_1) + (N_2 \times \bar{C}_2) \qquad (8.11a)$$

$$N \times \bar{S} = (N_1 \times \bar{S}_1) + (N_2 \times \bar{S}_2) \qquad (8.11b)$$

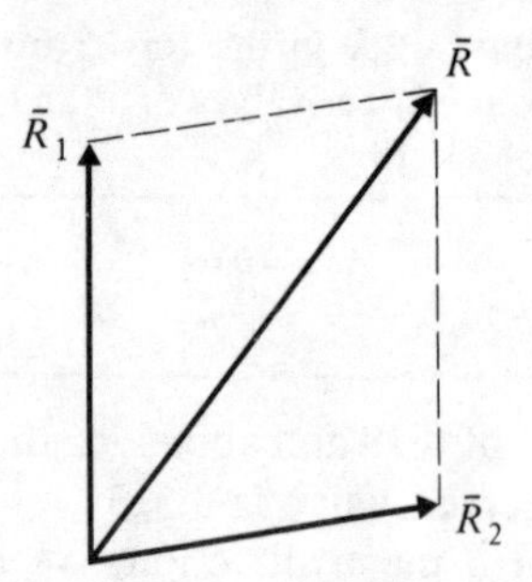

Figure 8.3 The parallelogram rule of vector addition: $\bar{R} = \bar{R}_1 + \bar{R}_2$.

from which:

$$\bar{R} = (\bar{C}^2 + \bar{S}^2) \tag{8.12}$$

The test for equality of mean directions assumes that the two samples come from populations of equal concentrations, so we should test for equality of concentrations first. Because the mathematics of the general test is so complicated, the practical test takes one of three forms (Mardia 1972, pp.158–62), depending on the value of $\bar{R}$. In all cases, the two samples should be approximately of equal size.

For $\bar{R}$ under 0.45, we can calculate a test statistic z:

$$z = 1.155 \times [g(\bar{R}_1) - g(\bar{R}_2)] / \surd[(N_1 - 4)^{-1} + (N_2 - 4)^{-1}] \tag{8.13}$$

where the function $g(\bar{R}_i) = \text{arc sin}\,(1.225 \times \bar{R}_i)$. (Take note that the angle is expressed in radians before taking the inverse sine (arc sin) and that x^{-1} is $1/x$, the reciprocal of x.)

Critical values of z appear in Table 5.3, the critical region containing the larger values.

When $\bar{R}$ is over 0.45 but under 0.70, z is given by:

$$z = [g(\bar{R}_1) - g(\bar{R}_2)] / \{0.893 \times \surd[(N_1 - 3)^{-1} + (N_2 - 3)^{-1}]\} \tag{8.14}$$

where the function $g(\bar{R}_i)$ is now defined:

$$g(\bar{R}_i) = \log_e[x + \surd(1 + x^2)]$$

$$x = (\bar{R}_i - 1.089)/0.258$$

Critical values of z are as above.

Lastly, when $\bar{R}$ is over 0.70, the test statistic F is given by:

$$F_{(N_1-1),(N_2-1)} = N_1 \times (1 - \bar{R}_1) \times (N_2 - 1) / [N_2 \times (1 - \bar{R}_2) \times (N_1 - 1)] \quad (8.15)$$

where F_{ν_1,ν_2} is Fisher's variance-ratio statistic. See Table 8.7 for critical values. In calculating F, sample '1' should be the smaller.

If we do not reject the null hypothesis that the samples are drawn from populations with equal concentration, then we can advance to a test of equality of mean directions. The test statistic t is given by:

$$t^2 = [1 + 3/(8\bar{k})] \times (N-2) \times (N_1 \times \bar{R}_1 + N_2 \times \bar{R}_2 - N \times \bar{R}) / (N - N_1 \times \bar{R}_1 - N_2 \times \bar{R}_2) \quad (8.16)$$

where $\bar{k}$ is the concentration parameter in the combined sample. Critical values of t are found in Table 5.5, with $\nu = N - 2$.

Table 8.6 Orientations (modulo 180°) of plant stems and goniatites in a bed (data kindly supplied by Dr W. B. Heptonstall).

Plant stems, N = 36		*Goniatites, N = 100*			
002	144	002	068	141	162
004	145	004	071	142	164
005	152	004	074	142	164
014	154	005	088	144	165
016	155	006	090	146	165
019	155	009	094	147	165
028	156	011	096	148	167
028	158	011	101	149	168
028	159	015	101	151	168
048	159	016	102	152	168
048	171	025	103	152	169
051	179	029	112	152	169
064		031	119	153	169
064		032	120	155	170
080		037	121	156	170
097		037	122	157	171
113		039	123	158	171
130		040	124	158	171
130		042	125	160	173
131		042	126	160	173
138		050	127	160	173
142		051	133	161	174
142		055	134	161	178
143		068	136	162	179
		068	141	162	179

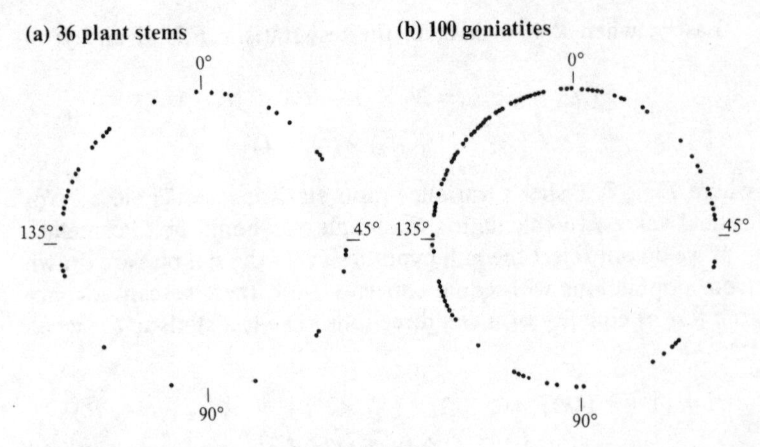

Figure 8.4 Two circular plots, modulo 180°.

Example. Table 8.6 and Figure 8.4 display measurements of the orientation of a sample of 36 plant stems and a sample of 100 goniatites in a bed. Both samples are of axial data, so Figure 8.4 is presented modulo 180° (i.e. once round the circle represents 180°). Using the tests above, we might compare the two samples.

First, the plots are reasonably unimodal, with no marked discontinuities in density, so an assumption that both are drawn from von Mises populations seems reasonable. Doubling the angles of Table 8.6 and applying Equations 8.6 a, b and 8.7 gives:

$$\bar{R}_p = 0.362$$

$$\bar{R}_g = 0.419$$

where the suffixes 'p' and 'g' refer to plants and goniatites respectively.

Applying the Rayleigh test of uniformity, Table 8.5, we find critical regions of size 0.01 containing values of $\bar{R}$ of 0.358 and above for the plants and 0.215 and above for the goniatites. Consequently, we may reject the null hypothesis that the two samples are drawn from uniform populations and proceed, realistically, with our analysis of the concentrations and mean directions. The concentrations are given approximately by Equation 8.9a as:

$$\bar{k}_p = 0.78$$

$$\bar{k}_g = 0.92$$

Consequently, it seems that the goniatites have a higher concentration (better 'preferred orientation') than the plant stems. To test this, we first require the mean resultant length of the combined sample, calculated by Equations 8.11a, b and 8.12:

$$\bar{R} = 0.404$$

Since this value is under 0.45, the test statistic z is calculated according to Equation 8.13:

$$z = -0.450$$

With a 'directional' alternative hypothesis (that k_g is greater then $\bar{k}_p$), we read from Table 5.3 a one-tail critical value of z of -1.64 for a critical region of size 0.05. Thus, we accept the null hypothesis of equality of the two concentration parameters.

Finally, with the mean directions of the plant stems and the goniatites calculated as 165° and 162° respectively (via Eq. 8.8 and remembering to halve the angles that result because the data are

Table 8.7 Critical values of Fisher's variance-ratio statistic F_{ν_1,ν_2}. ν_1 and ν_2 are both 'degrees of freedom' and equations should be arranged so that ν_1 is the smaller of the two. Critical regions contain values of F greater than the critical value. Upper lines: $a = 0.05$, lower lines: $a = 0.01$.

ν_1 \ ν_2	2	4	8	16	32	64	128
2	19.0	6.95	4.46	3.63	3.30	3.14	3.07
	99.2	18.1	8.65	6.24	5.34	4.95	4.78
4	19.3	6.39	3.84	3.01	2.67	2.52	2.44
	99.4	16.0	7.01	4.77	3.97	3.62	3.47
8	19.4	6.04	3.44	2.59	2.24	2.09	2.01
	99.5	14.8	6.03	3.89	3.13	2.80	2.65
16	19.4	5.84	3.20	2.33	1.97	1.80	1.72
	99.6	14.2	5.48	3.37	2.62	2.29	2.14
32	19.5	5.74	3.07	2.18	1.80	1.62	1.53
	99.6	13.9	5.18	3.08	2.32	1.98	1.83
64	19.5	5.68	3.00	2.10	1.71	1.51	1.41
	99.6	13.7	5.01	2.93	2.15	1.80	1.63
128	19.5	5.66	2.96	2.06	1.65	1.45	1.34
	99.6	13.6	4.93	2.84	2.06	1.70	1.51

modulo 180°), we should test for equality of mean directions. The test statistic t is given by Equation 8.16 (having calculated $\hat{k} = 0.88$ from Eq. 8.9a):

$$t = 0$$

(allowing for some numerical imprecision in the evaluation of the third set of brackets in the equation). Therefore, without further ado, accept the null hypothesis of equality of mean directions.

9 Three-dimensional orientations

Early in the careers of all of us, a compass/clinometer is thrust into our hands and a few, perhaps sketchy instructions given on its use. Rarely are any hints given on the quality of data that might be accumulated or on the quantity that should be accumulated. Consequently, studies are reported where sample sizes range from a dozen up to 10 000 or more, and in many minds it is unfortunate that the 'statistical' approach corresponds to the excess of the latter. Orientation data in three dimensions arise frequently and naturally in many geological studies, so this chapter contains a collection, by no means exhaustive, of methods in which statistical inference may be applied. Unfortunately, there are no nonparametric methods and the mathematics may become troublesome.

As shown in Section 1.2.2, orientation data may be either 'directions' or 'axes'. In three dimensions, the obvious choice for a graphical presentation is the stereographic or some similar projection, on which points represent lines in space or normals to planes. As with the circular plot, such a representation of the 'raw' data is the most convenient and honest form of display. To enhance the appearance of such diagrams a variety of methods of contouring the point density has been suggested but only one of these leads to a method of statistical hypothesis testing so I shall concentrate on this one only. First, a few notes follow on spherical data and its various distributions.

9.1 Spherical data and distributions

'Directions', with which both 'orientation' and 'sense' are associated, appear in geology in studies of palaeomagnetism, sedimentation palaeocurrents, etc. 'Axes' are not associated with a 'sense', and so geological examples might consist of lineations in metamorphic rocks, normals to layering or foliation, etc. In the case of circular data, the distinction between 'axes' and 'directions' tended to disappear under the application of the 'modulo transfor-

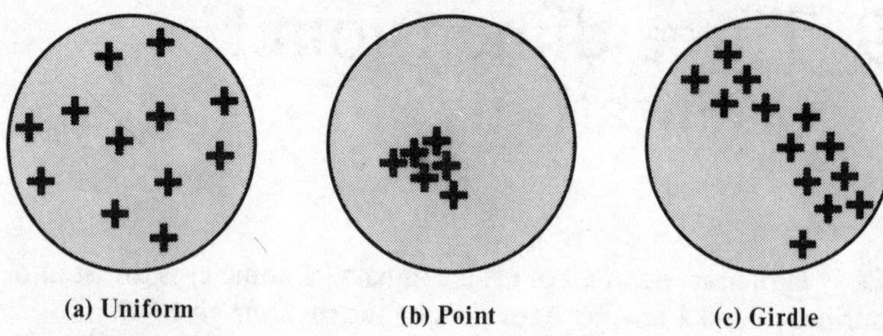

Figure 9.1 Three descriptive categories of spherical data represented on spherical plots. (a) Uniform: tendency for points to be uniformly distributed on an equal-area projection. (b) Point: tendency for points to cluster about a common axis. (c) Girdle: tendency for points to cluster about a great circle.

mation', but no corresponding manipulation exists for spherical data. By constructing projections of spherical data, three ill defined and merging categories of distribution of data, both 'directional' and 'axial', might be recognised, as in Figure 9.1.

The kinds of questions to which we are likely to want statistical answers include decisions as to whether or not a particular distribution differs significantly from uniform – if so, towards a 'point' or a 'girdle'? In the cases of 'points' and 'girdles', we need to find axes of rotational symmetry and appropriate measures of 'concentration' (or preferred orientation). For single samples, we might like to know 'confidence regions' for our axes of symmetry (i.e. regions of the stereogram into which the 'true' axis of symmetry falls with specified probability). Lastly, in the two-sample case, our usual question (whether or not the two samples represent the same population) has to be answered.

9.2 Point-density contours

Kamb (1959) has suggested a method by which density contours may be constructed so as to show concentrations which depart significantly from uniform. Figure 9.2a shows an equal-area projection containing a small 'counting circle' which has a fractional area H relative to the projection circle, so that H may take any value from 0 to 1. If the projection contains N data points altogether, and we suggest that these N points are drawn from a uniform population, then we expect $H \times N$ points inside the counting circle and $(1 - H) \times N$ points outside. If R is the number of points observed in the counting circle and if R is significantly

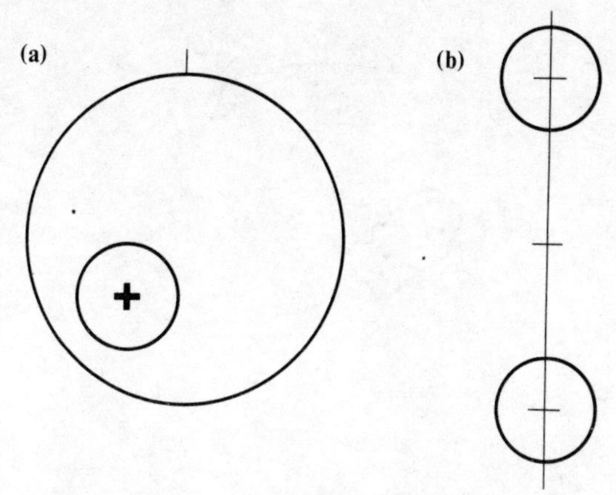

Figure 9.2 'Counting circles' for point-density measurement. See text for full explanation.

greater than the expected number $H \times N$, then we reject the null hypothesis of uniformity and accept an alternative that the array of points contains one or more significant clusters. The sampling distribution of R is binomial so that probabilities of various values of R can be calculated by Equation 2.1. If we make $H = 1/N$, then we find, for N over 10, that values of R of 4 and over (or 5 and over) have probabilities of occurrence of 0.01 or less (or 0.001 or less), respectively, under the null hypothesis, irrespective of the sample size. Consequently, the practical test can take this form:

(1) Plot the data on an equal-area projection.
(2) Measure the radius r of the projection circle and calculate the radius $r/\sqrt{(N)}$ of the 'counting circle' (i.e. making its area $1/N$ of that of the projection circle).
(3) *Either* draw a circle of this radius about every data point *or* search the area of the projection with the counting circle, looking for concentrations where four, five or more points (depending on our chosen size of critical region) fall within the counting circle. Near the circumference of the projection, both alternatives are facilitated by a double counting circle, drawn on tracing paper (as in Fig. 9.2b), where the distance between the centres of the counting circles is $2 \times r$. The line joining the centres of the circles is placed along a diameter of the projection circle and the point density is measured by counting and adding the contents of both circles.

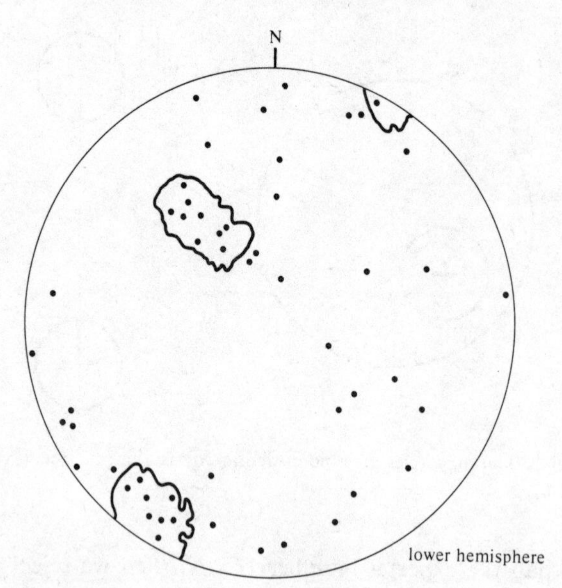

Figure 9.3 Edinburgh Castle Rock, or rather an equal-area, lower hemisphere projection of downward-pointing normals to 50 joint surfaces on the south side of the Rock. Irregular-shaped loops enclose areas in which point density departs significantly from uniform.

(4) Draw contours separating areas of point density of four, five or over, per counting circle, from areas of point density less than four or five.
(5) For the 'high-density' areas, reject the null hypothesis of uniform point density with a critical region of size 0.01 or 0.001 or less, as appropriate.

An example of a diagram contoured in this way is shown in Figure 9.3, an equal-area, lower hemisphere projection of the downward-pointing normals to 50 joints on the south side of Edinburgh Castle Rock. There appear to be two concentrations where density is significantly above uniform. Strictly, the 'counting circle' used in this method should be an ellipse, the area of which remains constant, but with eccentricity dependent on position. However, such refinements are pedantic for practical use.

9.3 Numerical representation of spherical data

For patterns that depart from 'uniform' towards either 'point' or 'girdle' distributions, we are likely to find the theoretical distributions of Fisher and Bingham of use – these are introduced below. However, before we can proceed with the manipulation of orientation data as it is normally gathered in geological studies, we have to introduce an alternative method of numerical representation. Nearly all orientation data in three dimensions can be simplified either to **linear** or **planar**. That is, the physical properties whose orientations we wish to specify can be represented geometrically either by lines or by planes. For convenience of presentation on the spherical projection, we usually choose to specify the orientation of a plane by the orientation of its normal. Furthermore, in the case of lines, we measure an **angle of inclination** (or 'angle of plunge'), positive vertically downwards (and negative upwards) from the horizontal, and an **azimuth**, the angle measured clockwise from the north to the projection of the line on to the horizontal plane. For planar data, we usually specify orientations by measuring an 'angle of dip' vertically downwards from the horizontal to the line of maximum slope in the plane, and an azimuth (or 'direction of dip'), clockwise from north to the projection of the line of maximum slope on to the horizontal plane. However, such numerical representations are unsatisfactory for direct statistical manipulations.

Figure 9.4 shows a basis for a suitable transformation of our conventional geological measurements. First, we need three mutually perpendicular co-ordinate reference axes whose positive ends can conveniently point north, east and down. In such a co-

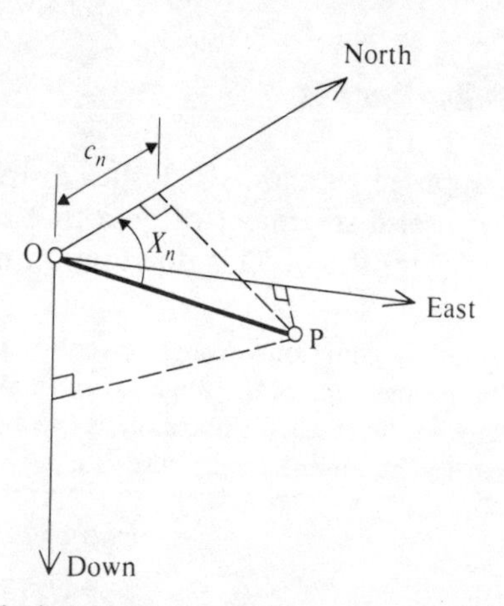

Figure 9.4 A geological direction is represented by a line of unit length OP. For convenience of computation, the direction is specified by quoting the lengths of the perpendicular projections of OP on to reference axes that point north, east and down. These lengths are called 'direction cosines'. See text for full explanation.

ordinate system, let any line OP represent a direction of interest, an item of linear data or a normal to an item of planar data. Let the length of OP be one unit, so that it becomes a **unit vector**, and let perpendiculars be dropped from P on to each of the three reference axes (i.e. project OP on to each of the three reference axes in turn). If X_n, X_e and X_d are the angles made by OP with the reference axes pointing north, east and down respectively (largely omitted from Fig. 9.4 for clarity), then the lengths of the projections of vector OP on to the reference axes are:

$$c_n = \cos X_n$$
$$c_e = \cos X_e \qquad (9.1)$$
$$c_d = \cos X_d$$

These quantities are known as **direction cosines** and may be calculated from conventional geological data as in Table 9.1.

Alternatively, since a high order of numerical precision is not generally required in this work, alert readers will see that a special net for the spherical projection can be drawn, containing three families of small circles concentric about north, east and down respectively. These small circles can be graduated not in angles, but in the cosines of the angles, so that the three direction cosines corresponding to a point on the projection can be read off directly to a precision of about three significant figures.

9.4 The Fisher distribution

For directions as opposed to axes, the Fisher distribution on the sphere provides for useful treatment of data that tend towards a point concentration (Fig. 9.1b). This distribution may be looked

Table 9.1 Transformation equations from conventional geological coordinates to direction cosines. *inc* is the angle of inclination, 'plunge' in the case of a line, 'dip' in the case of a plane. *az* is the azimuth measured clockwise from north to the plunge or dip direction.

Lines	*Normals to planes*
$c_n = \cos(inc) \times \cos(az)$	$c_n = -\sin(inc) \times \cos(az)$
$c_e = \cos(inc) \times \sin(az)$	$c_e = -\sin(inc) \times \sin(az)$
$c_d = \sin(inc)$	$c_d = \cos(inc)$

upon as the analogue on the sphere of the von Mises distribution on the circle and has a probability density function, quoted for the sake of completeness only, given by:

$$f(\mathbf{c}) = k \times \exp(k \times \mathbf{c}.\mathbf{l})/[2 \times \sinh(k)] \qquad (9.2)$$

where the *parameters* are k, a *concentration* and $\mathbf{l}$, a unit vector specifying the *mean direction*. $\mathbf{c}$ is a unit vector drawn at random from the population, $\mathbf{c}.\mathbf{l}$ is the 'scalar product' (Macbeath 1964, or any book on elementary vector algebra), and $\sinh(k)$ is the hyperbolic sine function whose argument is k.

9.4.1 Parameters and estimation

As with the von Mises distribution, the concentration k ranges from zero for a uniform distribution to plus infinity for a distribution in which all vectors point the same way. It is thus a dispersion type parameter and can be used as a measure of preferred orientation. The unit vector $\mathbf{l}$, specifying the mean direction, can also be represented by its direction cosines l_n, l_e and l_d which help to locate it on a spherical projection. For a sample which is thought to be drawn from a Fisher population, the mean direction and concentration parameter may be estimated as follows:

(1) Define

$$\bar{c}_n = \Sigma(\cos X_n)/N$$
$$= \Sigma(c_n)/N \qquad (9.3)$$

with similar expressions for $\bar{c}_e$ and $\bar{c}_d$, the summation (Σ) covering the whole sample, i.e. simply calculate arithmetic means of the 'north', 'east' and 'down' direction cosines respectively.

(2) Calculate the mean resultant length $\bar{R}$:

$$\bar{R} = \sqrt{(\bar{c}_n{}^2 + \bar{c}_e{}^2 + \bar{c}_d{}^2)} \qquad (9.4)$$

(3) The direction cosines of the estimated mean direction are given by:

$$\bar{l}_n = \bar{c}_n/\bar{R} \qquad \bar{l}_e = \bar{c}_e/\bar{R} \qquad \bar{l}_d = \bar{c}_d = /\bar{R} \qquad (9.5)$$

from which the mean direction may be found.

(4) For estimation of $\bar{k}$, the concentration, calculation is generally difficult but the approximation:

$$\bar{k} = 1/(1 - \bar{R}) \tag{9.6}$$

is sufficient if $\bar{R}$ is above about 0.65.

Example. Table 9.2 contains observations of the orientation of bedding and cleavage at 15 localities spread over 1 km of Silurian greywackes in South Scotland. Angles of dip in excess of 90° mean the bedding or cleavage is 'overturned', cleavage 'facing' in the direction in which it intersects younger beds. Figure 9.5 presents the data graphically, points representing downward-directed normals in a lower hemisphere projection.

For the bedding orientation measurements, Equation 9.3 gives the arithmetic means of the direction cosines as:

$$\bar{c}_n = -11.729/15 = -0.782$$

$$\bar{c}_e = +7.211/15 = +0.481$$

$$\bar{c}_d = -4.764/15 = -0.318$$

Table 9.2 Observations of the orientations of bedding and cleavage at localities in southern Scotland.

Locality	*Bedding (dip/ azimuth)*	*Direction cosines of normals on older side* c_n	c_e	c_d	*Cleavage (dip/ azimuth)*
1	119/335	−0.733	+0.370	−0.485	108/357
2	119/331	−0.765	+0.424	−0.485	109/341
3	105/343	−0.924	+0.282	−0.259	110/330
4	114/325	−0.748	+0.524	−0.407	105/334
5	87/321	−0.776	+0.628	+0.052	84/345
6	107/326	−0.793	+0.535	−0.292	107/341
7	102/322	−0.771	+0.602	−0.208	111/343
8	109/326	−0.784	+0.529	−0.326	104/339
9	118/331	−0.772	+0.428	−0.469	119/345
10	125/333	−0.730	+0.372	−0.574	119/347
11	122/327	−0.711	+0.462	−0.530	109/348
12	81/337	−0.909	+0.386	+0.156	80/341
13	110/313	−0.641	+0.687	−0.342	
14	100/332	−0.870	+0.462	−0.174	
15	115/325	−0.742	+0.520	−0.423	124/348
totals		−11.729	+7.211	−4.764	

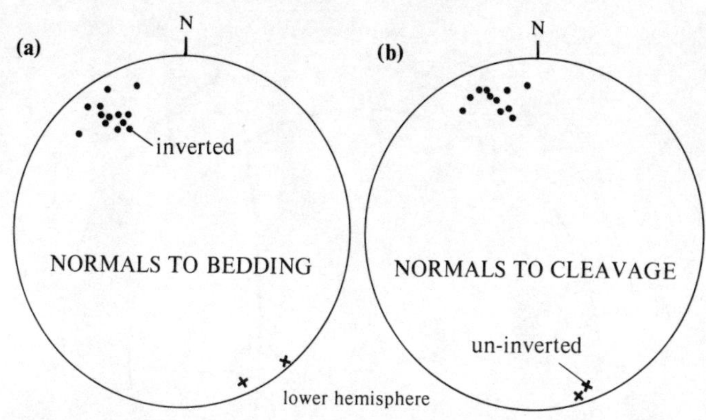

Figure 9.5 Downward-pointing normals to bedding (a) and cleavage (b) in a Silurian greywacke formation in southern Scotland. Full circles represent normals to inverted beds or downward-facing cleavage, crosses represent normals to right-way-up beds or upward-facing cleavage.

The mean resultant length $\bar{R}$ is the square root of the sum of the squares of these three numbers (Eq. 9.4):

$$\bar{R} = 0.971$$

so that the direction cosines of the estimated mean direction are given by Equations 9.5:

$$\bar{l}_{n} = -0.782/0.971 = -0.805 \ (143°)$$

$$\bar{l}_{e} = +0.481/0.971 = +0.495 \ (60°)$$

$$\bar{l}_{d} = -0.318/0.971 = -0.327 \ (109°)$$

The angles corresponding to these direction cosines are given in brackets. These angles may be used to locate the estimated mean direction on the spherical projection by constructing small circles of radius 143°, 60° and 109° about the 'north', 'east' and 'down' points respectively (Fig. 9.6). Alternatively, equations in Table 9.1 may be rearranged and solved to give:

$$\begin{aligned} \textit{dip} &= \text{arc cos } (\bar{c}_{d}) = 109° \\ \textit{azimuth} &= \text{arc cos } [\bar{l}_{n}/-\sin(\text{dip})] = 33° \text{ or } 327° \\ &= \text{arc sin } [\bar{l}_{e}/-\sin(\text{dip})] = 213° \text{ or } 327° \end{aligned}$$

(remember that inverse trigonometrical functions have more than one result)

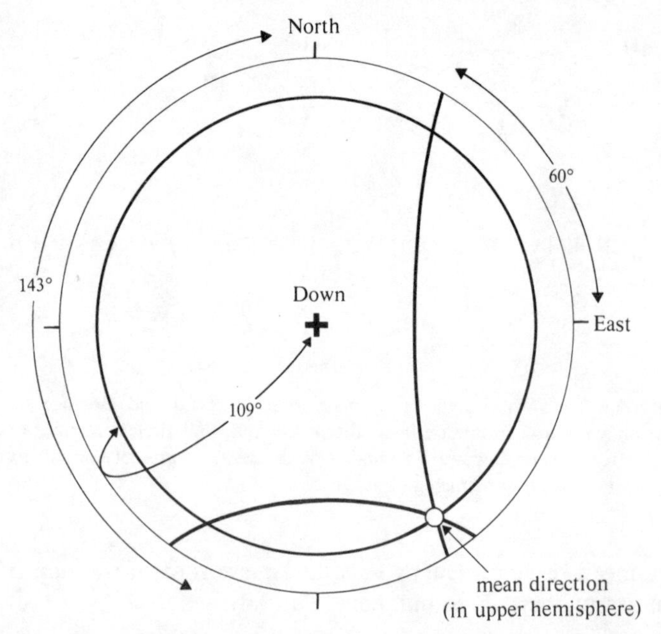

Figure 9.6 The construction on the projection needed to locate the mean direction of bedding of the data given in Table 9.2. Small circles of radii 143°, 60° and 109° are constructed about centres at 'north', 'east' and 'down' respectively. They intersect at the required mean direction. Note that because the angle subtended with the 'down' axis exceeds 90°, the resulting mean direction is plotted in the upper hemisphere.

so that the mean direction of bedding is 109° of dip towards azimuth 327°. Finally, an estimate of the concentration $\bar{k}$ is given by Equation 9.6: $\bar{k} = 35$.

For the cleavage data, similar calculations give $\bar{R} = 0.972$ with mean direction dipping at 107° towards azimuth 343° and concentration $\bar{k} = 36$.

9.4.2 Confidence intervals and cones

Having estimated these parameters, it is useful to be able to quote confidence intervals. Here, as was the case with the von Mises distribution, confidence intervals for the concentration parameter are difficult to calculate and are of subsidiary practical application in any case. However, a **cone of confidence** is easily calculated for the mean direction and is especially useful in, for example, palaeomagnetic studies as well as those reported above. Because the Fisher distribution has rotational symmetry about its mean direction, the confidence interval for this mean direction also has rotational symmetry and can be represented geometrically by a circular cone co-axial with the mean direction. On the spherical projection, the confidence cone appears as a small circle with the mean direction at its centre (e.g. Fig. 9.7). If the product $N \times \bar{R} \times \bar{k}$ is over 3, then the semi-apical angle of the cone d is given by:

$$d = \text{arc cos } [1 + \log_e a/(N \times \bar{R} \times \bar{k})] \qquad (9.7)$$

where $a = 0.01$ for 0.99 confidence and 0.05 for 0.95 confidence. Take care with the priorities of arithmetical operators when evaluating this expression (see the note at the beginning of the book).

For samples in Table 9.2, the semi-apical angles for 0.99 confidence cones are $d = 8°$ in both cases. Consequently, small circles of this radius may be contructed with estimated mean directions as centres as shown in Figure 9.7. There is thus a probability of 0.99 that the 'true' mean directions are to be found inside these small circles.

9.4.3 Two-sample tests

Two-sample tests of concentration parameters and mean directions may be constructed for samples drawn from Fisher populations and are analogous in many ways to those for von Mises populations

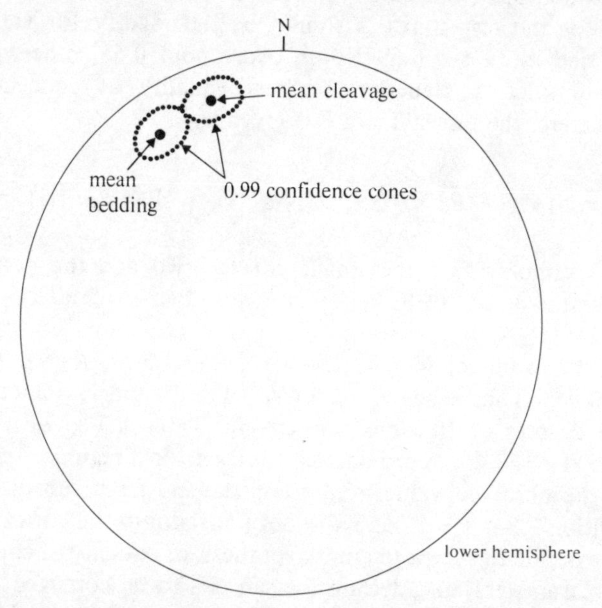

Figure 9.7 0.99 confidence cones for the data of Table 9.2. See text for constructional details. Equal-area projection.

(Section 8.3.2). As with circular distributions, the test for differences between mean directions assumes equality of concentrations so the latter condition should be tested first. Tests of both concentration and mean direction depend on calculation of mean resultant lengths of combined samples, and the method of doing this is exactly analogous to Equations 8.11a,b and 8.12 for the circular case. In the current example, we have:

sums of direction cosines:	c_n	c_e	c_d
bedding (15 specimens):	−11.729	+7.211	−4.764
cleavage (13 specimens):	−11.553	+3.533	−3.704
totals (28 specimens):	−23.282	+10.744	−8.468
means of direction cosines (i.e. totals/28):	−0.832	+0.384	−0.302

so that the combined mean resultant length is given by:

$$\bar{R} = \surd[(-0.832)^2 + (+0.384)^2 + (-0.302)^2] = 0.965$$

As in the circular case, the two-sample tests are based on approximations (Mardia 1972, p.267), but the following expressions for the test statistics should be satisfactory for values of combined mean resultant length over about 0.65, otherwise the computations are lengthy. To test equality of concentration parameters, the test statistic F is given by:

$$F_{\nu_1,\nu_2} = N_1 \times (1 - \bar{R}) \times (N_2 - 1)/[N_2 \times (1 - \bar{R}) \times (N_1 - 1)] \qquad (9.8)$$

where sample '1' is the smaller of the two and the degrees of freedom ν_1 and ν_2 of Fisher's F are given by: $\nu_1 = 2 \times (N_1 - 1)$ and $\nu_2 = 2 \times (N_2 - 1)$.

In the example, $N_1 = 13$, $N_2 = 15$, $\bar{R}_1 = 0.971$, $\bar{R}_2 = 0.972$ and $\bar{R} = 0.965$. Thus F has $\nu_1 = 2 \times (N_1 - 1) = 24$ and $\nu_2 = 2 \times (N_2 - 1) = 28$ degrees of freedom respectively. Table 8.7 gives a critical value of $F = 2.66$, approximately, for a critical region of size 0.01 and the observed value of the test statistic from substitution in Equation 9.8 is $F = 1.05$. This does not fall into the critical region so we decide to accept the null hypothesis of equality of concentration parameters, and declare the samples to be homogeneous.

The test statistic for the two-sample test of mean directions

(Mardia 1972, p. 263) is also F, calculated by:

$$F_{2,2\times N-2} = (N-1)\times(N_1\times\bar{R}_1 + N_2\times\bar{R}_2 - N\times\bar{R})/(N - N\times\bar{R}) \tag{9.9}$$

where N is the combined sample size.

For the example, F has '2' and '54' degrees of freedom so the critical value of a region of size 0.01 is approximately 5. Substitution in Equation 9.9 yields an observed value of the test statistic of $F = 5.04$. This is very close to our approximate critical value. We are certainly well inside a critical region of size 0.05 (critical value about 3.2) but reference to a fuller table appears necessary. However, we may note that $F_{2,\nu}$ is given exactly by

$$F_{2,\nu} = \frac{\nu}{2}\times(a^{-2/\nu} - 1) \tag{9.10}$$

where a is the chosen size of the critical region (note $a^{-f} = 1/a^{f}$).

Substitution in Equation 9.10 gives an accurate critical value of the test statistic of $F = 5.02$, leading to rejection of the null hypothesis of equality of mean directions with a critical region of size 0.01.

9.5 The Bingham distribution

9.5.1 Introduction

The Bingham distribution (Bingham 1964) is a general distribution for axial data and, according to the values of its parameters, it may be used to give a numerical summary of any of the three categories of Figure 9.1 or any gradation between them that maintains an orthorhombic symmetry (i.e. with three mutually perpendicular planes of symmetry). Thus, a density-contoured stereographic projection may have the appearance of Figure 9.8. If such a distribution is thought of as having three mutually perpendicular planes of symmetry, then these planes intersect along three mutually perpendicular axes. If these three **principal axes** can be located, and some estimate of the point density in their directions can be made, then we have a potentially extremely useful numerical description of the overall distribution. However, be warned that what follows may be hard work! Use of the Bingham distribution relies on a knowledge of matrix algebra so a short glossary is provided in Appendix A, otherwise consult some elementary text such as Hall (1963).

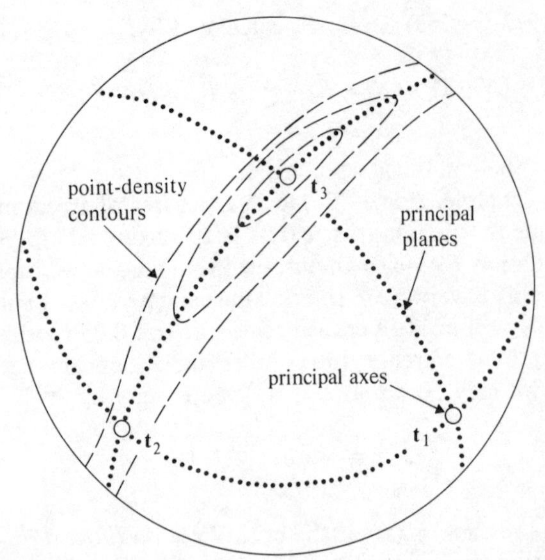

Figure 9.8 Some of the features of symmetry of the Bingham distribution discussed in the text.

The Bingham distribution is formally defined by its probability density function as follows, but again, only for reference:

$$f(\mathbf{c}) = [4 \times \pi \times d(\mathbf{K})]^{-1} \times \exp[K_1 \times (\mathbf{c}.\mathbf{t}_1)^2 + K_2 \times (\mathbf{c}.\mathbf{t}_2)^2 + K_3 \times (\mathbf{c}.\mathbf{t}_3)^2] \qquad (9.11)$$

where **c** is a vector representing an axis drawn at random from the population, and the *parameters* of the population are given by a diagonal matrix of *concentrations* **K** (with diagonal elements K_1, K_2 and K_3) and three mutually perpendicular *principal axes* represented by the vectors $\mathbf{t}_1$, $\mathbf{t}_2$ and $\mathbf{t}_3$. (Check Appendix A for unfamiliar terms.)

Any sample drawn from a Bingham population can be described by six numbers calculated directly from the orientations of the sample axial data. These six numbers can be arranged in a symmetrical matrix **T** of order (3,3). Associated with **T** are three eigenvectors which estimate the directions of the principal axes ($\mathbf{t}_1$, $\mathbf{t}_2$ and $\mathbf{t}_3$) of the parent population and then associated with each of the three eigenvectors is an eigenvalue τ proportional to point density in that direction. By convention, the three eigenvalues are labelled τ_1, τ_2 and τ_3 in order of increasing magnitude. The three eigenvalues may be used to calculate the matrix of concentration parameters **K** but this seems rarely necessary in practice, it being sufficient to note

Table 9.3 Categorisation of Bingham distributions.

Category	*Eigenvalues* τ
uniform	approximately equal
point (or 'bipolar)	τ_1 and τ_2 small and equal, τ_3 large
girdle	τ_1 small, τ_2 and τ_3 large and equal

that the relative magnitudes of the three eigenvalues are approximately similar to the relative magnitudes of the three corresponding elements of **K**. Because of this last property, the relative magnitudes of the eigenvalues can be used to categorise distributions as in Table 9.3

9.5.2 *Parameters and estimation*

The calculations associated with the use of the Bingham distribution follow stages exactly analogous to those associated with the Fisher distribution, with a time or labour penalty about three times greater because of the extra arithmetic involved. There is no great difficulty in doing these 'by hand' with a simple calculator, but a

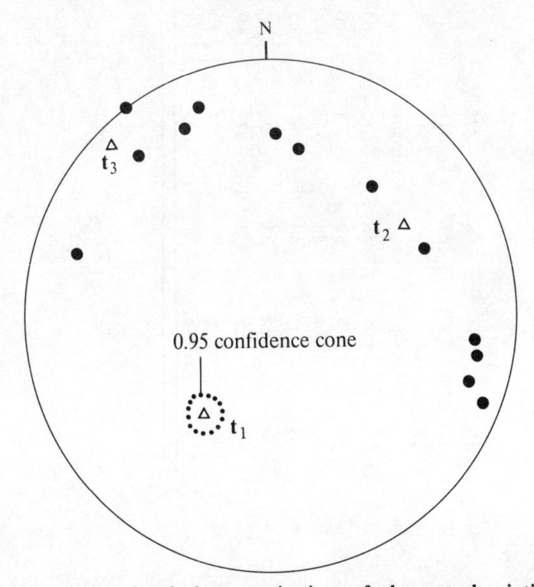

Figure 9.9 The lower hemisphere projection of downward-pointing normals (marked by full circles) to bedding in a Silurian greywacke formation at Eyemouth, Berwickshire, Scotland. The principal axes of the distribution are marked by open triangles. The principal axis corresponding to the smallest eigenvalue, τ_1, is labelled $\mathbf{t}_1$ and is surrounded by a 0.95 confidence cone. This axis represents the best estimate of the fold axial direction.

Table 9.4 Calculation of eigenvectors and eigenvalues in a sample of axial data.

		X_n	cos	X_e	cos	X_d	cos	*Check*			*Notes*
locality	1	19	+946	106	−276	80	+174	1.001			(1)
on the	2	31	+857	121	−515	90	000	1.000			
synform	3	38	+788	124	−559	75	+259	1.001			
	4	26	+899	111	−358	74	+276	1.013			
	5	25	+906	86	+070	65	+423	1.005			
	6	32	+848	80	+174	60	+500	0.999			
	7	50	+643	58	+530	56	+559	1.006			
	8	70	+342	40	+766	57	+545	1.001			
	9	100	−174	23	+921	70	+342	0.995			
	10	97	−122	22	+927	69	+358	1.002			
	11	106	−276	27	+891	70	+342	0.987			
	12	111	−358	27	+891	75	+259	0.989			
							$\overline{\mathbf{T}}$ =	+5.378	−1.616	+1.680	(2)
								−1.616	+4.980	+1.721	
								+1.680	+1.721	+1.641	
						$\overline{\mathbf{T}}$ −	12**I** =	−6.622	−1.616	+1.680	(3)
								−1.616	−7.020	+1.721	
								+1.680	+1.721	−10.359	

iteration	$\mathbf{t}_1 \doteqdot$	112	−375	123	−545	41	+755	+4.632	+5.731	−9.389	(4)
number	1	113	−388	119	−480	38	+787	+4.667	+5.351	−9.630	
	2	113	−390	117	−447	36	+805	+4.657	+5.154	−9.763	
	3	113	−388	115	−430	36	+814	+4.632	+5.047	−9.824	
	$\mathbf{t}_1 =$	113	−387	115	−421	35	+820	−0.023	−0.060	−0.029	
iteration	$\mathbf{t}_3 \doteqdot$	54	+588	54	+588	56	+599	+3.151	+2.940	+2.917	(5)
number	1	53	+606	56	+565	56	+561	+3.289	+2.800	+2.911	
	2	51	+631	58	+537	56	+559	+3.465	+2.617	+2.902	
	3	48	+664	60	+501	56	+556	+3.695	+2.379	+2.890	
	4	45	+702	63	+452	57	+549	+3.967	+2.061	+2.858	
		42	+748	67	+388	57	+539				
iteration	$\mathbf{t}_3 \doteqdot$	36	+809	126	−588	85	+087	+5.447	−4.086	+0.490	
number	1	37	+798	127	−598	86	+072	+5.379	−4.144	+0.430	
	2	38	+791	128	−609	86	+063	+5.344	−4.203	+0.384	
	$\mathbf{t}_3 =$	38	+785	128	−617	87	+056	+5.313	−4.245	+0.349	
	$\mathbf{t}_2 =$	61	+482	48	+665	55	+569	+2.473	+3.512	+2.888	(6)

Notes

(1) Form matrix of normals to bedding. Enter direction cosines.

(2) Form matrix **T**. Check diagonal elements sum to $N = 12$.

(3) Subtract 12**I**.

(4) Find minimum eigenvalue τ_1, by starting with an approximate solution for $\mathbf{t}_1$. Operate on $\overline{\mathbf{T}} - 12 \times \mathbf{I}$.

(5) Find maximum eigenvalue τ_3, by operating on $\overline{\mathbf{T}}$ with an approximate solution for $\mathbf{t}_3$. Note rapid divergence. Start again with a better approximation.

(6) Find $\mathbf{t}_2$ perpendicular to the other two vectors.

programmable machine speeds things up a lot. In outline, the first stage is the trigonometrical manipulation of the data to derive the matrix **T** (cf. summation of sine and cosine formulae in connection with the Fisher distribution). The second stage is to calculate the orientations of the principal axes (cf. mean direction) and the third is to derive the eigenvalues (cf. mean resultant length) from which concentration parameters may be found if needed. The method outlined below combines the second and third stages for convenience and is illustrated by a worked example, based on the distribution of the downward-pointing normals to bedding in synformally folded Palaeozoic rocks near Eyemouth, Berwickshire, shown in Figure 9.9. The stages of the calculation outlined below may be followed in Table 9.4.

(1) Derive the *direction cosines* (see Section 9.3) associated with the data points. Together with the corresponding angles X_n, X_e and X_d, these are shown in the top of the table. Decimal points have been omitted and only the first three decimal figures presented for economy of space and effort. As a check at each locality, the sum of the squares of the three direction cosines should be near unity.

(2) Form the matrix **T**. If the matrix of direction cosines is **D**, then **T** is obtained by pre-multiplication of **D** by its transpose **D**′ (see Appendix A) so that $\mathbf{T} = \mathbf{D}' \times \mathbf{D}$. If you are a newcomer to matrix methods, do not be deterred; the evaluation involves no more than the adding together of squares and products of cosines. As a partial check, the diagonal elements $(+5.378 + 4.980 + 1.641 = 11.999)$ should sum to near 12, the sample size being N.

(3) Form the matrix $\mathbf{T} - N \times \mathbf{I}$. The sample size $N = 12$, and **I** is the unit matrix of order 3. Thus simply subtract 12 from each of the diagonal elements of **T**. The matrix so formed is suitable for finding the eigenvector $\mathbf{t}_1$ and the corresponding minimum eigenvalue τ_1.

(4) To find the eigenvector $\mathbf{t}_1$, start with an approximation for $\mathbf{t}_1$. A suitable approximation is easily found from the spherical projection of the data by locating 'by eye' the pole to the 'best-fit' great circle. In this example, the values $X_n = 112°$, $X_e = 123°$ and $X_d = 41°$ were used. By following the procedure for the calculation of eigenvalues and eigenvectors outlined in Appendix A using these figures, four iterations were found sufficient for the approximation to converge to a satisfactorily accurate $\mathbf{t}_1$.

(5) To find the eigenvector $\mathbf{t}_3$ and the corresponding eigenvalue τ_3, operate on **T** itself, as above, starting with an approximation for $\mathbf{t}_3$ obtained from the projection at a position where the point density appears to be a maximum. In the example, the first approximation proved unsuitable because successive iterations diverged from the initial guess. A fresh start with a new approximation was much more satisfactory.

(6) To find the eigenvector $\mathbf{t}_2$ simply plot the solutions for $\mathbf{t}_1$ and $\mathbf{t}_3$ on the projection and find that direction that is perpendicular to them both. The corresponding eigenvalue τ_3 is found as outlined in Appendix A.

The eigenvalues calculated in this example are: $\tau_1 = 0.0705$ $\tau_2 = 5.176$ and $\tau_3 = 6.811$. As a final check, these should sum approximately to the sample size N.

There are three uses to which these estimates of population eigenvectors and eigenvalues may be put. First, comparing the eigenvalues obtained with entries in Table 9.3 supplies the trivial confirmation that the sample is probably drawn from a girdle population. Secondly, the eigenvectors may be added to the projection (Fig. 9.9) and, if required, the 'best-fit' great circle may be drawn through $\mathbf{t}_2$ and $\mathbf{t}_3$. Thirdly, and potentially most useful, a cone of confidence can be constructed about $\mathbf{t}_1$, the pole of this 'best-fit' great circle.

9.5.3 *Confidence cones*

Because the Bingham distribution generally has orthorhombic rather than rotational symmetry, cones of confidence are elliptical in section rather than circular, so that the semi-apical angle d varies with meridian m (Fig. 9.10). In the case of a cone about the $\mathbf{t}_1$ principal axis, the semi-apical angle is given by:

cone of confidence about $\mathbf{t}_1$:

$$\sin^2 d = \tau_1 \times (a^{-2/(N-2)} - 1)/(\tau_2 \times \cos^2 m + \tau_3 \times \sin^2 m - \tau_1) \tag{9.12}$$

where $(1 - a)$ is the required confidence level. So the values of d may be calculated for convenient values of m.

The angles d and m may be related to $\mathbf{t}_1$, $\mathbf{t}_2$ and $\mathbf{t}_3$ as in Figure 9.10 which serves as a basis for constructing the cones on the spherical projection. In the example, a 0.95 cone of confidence has

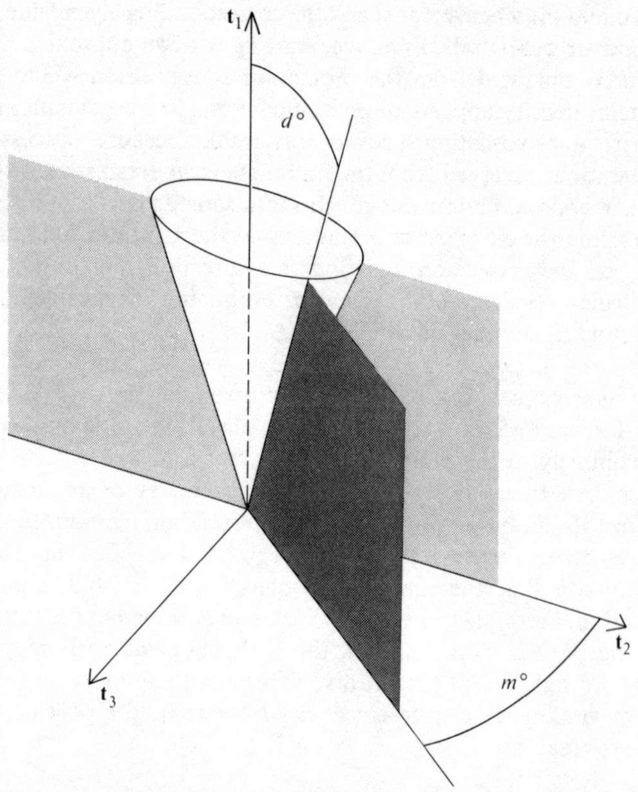

Figure 9.10 The basis of the method of constructing elliptical confidence cones derived from application of the Bingham distribution. The light-toned plane represents the $\mathbf{t}_1 \sim \mathbf{t}_2$ principal plane. The meridional angle m is measured from $\mathbf{t}_2$ towards $\mathbf{t}_3$ to define the position of the dark-toned plane in which the semi-apical angle d is measured from $\mathbf{t}_1$. The dark-toned plane contains the $\mathbf{t}_1$ axis. For cones about $\mathbf{t}_3$, interchange $\mathbf{t}_1$ and $\mathbf{t}_3$.

been constructed (Fig. 9.9). Because τ_2 and τ_3 are so close, this cone is nearly circular with semi-apical angle d near to 6°.

For a 'point' or 'bipolar' distribution, a cone of confidence about $\mathbf{t}_3$, the principal axis corresponding to the largest eigenvalue and therefore to the highest density of points, is likely to be more useful. This cone is given by:

cone of confidence about $\mathbf{t}_3$:

$$\sin^2 d = -\tau_3 \times (a^{-2/(N-2)} - 1)/(\tau_1 \times \cos^2 m + \tau_2 \times \sin^2 m - \tau_3) \tag{9.13}$$

where $(1 - a)$ is the required confidence level. The cone axis is now $\mathbf{t}_3$ and m is measured from $\mathbf{t}_2$ towards $\mathbf{t}_1$.

Thus armed with a means of calculating confidence cones, the construction of one-sample tests of naturally occuring distributions against theoretical predictions becomes elementary. Two-sample tests may also be constructed, as Mardia (1972) demonstrates, thus allowing a wide variety of hypotheses concerning three-dimensional axial data to be tested in those situations where the null hypothesis suggests the two samples are drawn from the same parent population. More troublesome are those instances where the two samples can be supposed to be drawn from Bingham populations, but populations of manifestly different characters between which it is only required to compare certain parameters. I have in mind the situation in which it is required to test whether or not a sample of linear structures can be considered as lying parallel to a girdle axis derived from a second sample; for example, parallelism of lineations with major fold hinges. This is an area in which further research would be useful.

10 Hypotheses, samples and decisions

I have made the assumption that the user of statistical methods 'on site' does so in order to supplement other tools and facilities and that statistical methods are not applied as an end in themselves. To a certain extent advance and consolidation in many scientific studies is a kind of cyclical process and earlier ideas are revised as newer data become available – so it is with elementary geological studies covering only a small part of the discipline. If we take one of the principal aims of a study to be the preparation of an objective statement, then for the purposes of the discussion below, the procedures can also be broken down into cyclically arranged stages as in Figure 10.1.

10.1 A possible route to objectivity

The first stage of the procedure consists of perceiving an area of study and gaining some familiarity with it (i.e. getting interested). Thus, initiative or edict will introduce a problem or project and the initial effort is one of accumulating primary information from sources such as literature, verbal communication, pilot survey, etc. to provide a preliminary description. This description should then serve as a source of stimulation and definition for further lines of investigation, guided or otherwise.

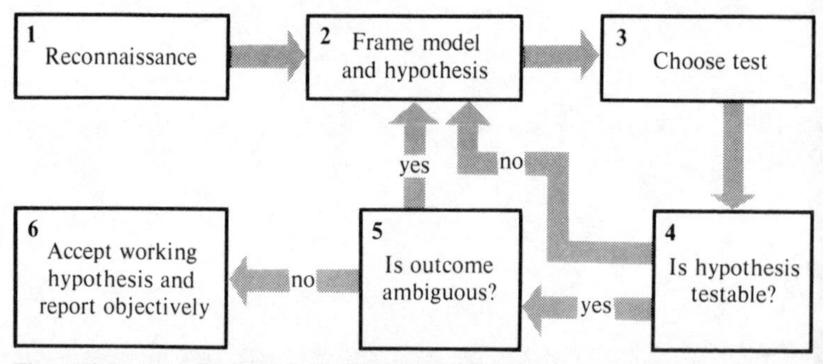

Figure 10.1 A cyclical approach to statistical problems.

Stage 2 requires the worker to tidy up the intuitive thoughts that should have arisen earlier, in short, to change 'bright ideas' into 'sound hypotheses'. Here, some combination of ingenuity and previous (or received) experience are likely to be the best aids. To elevate subjective intuition to statistically testable hypothesis, we must define a statistical model; that is, identify and enumerate a group of related concepts which we hope will lead to a quantified description of some or all of our study area. Stated simply, this model is likely to consist of an identification and count of the measurable variables together with some assertions about the populations they purport to describe. For example, a study of the size, shape and composition of beach pebbles might be based on three variables: 'length', 'roundness' and 'rock type'. Scales and modes of measurement of each will have to be defined. Any assumptions made concerning their distributions in the populations will be listed. A further essential component of the statistical model will be a careful description of the method of sampling the population: indeed, sampling is such an important topic that I postpone fuller discussion to the next section. Finally, the range of statistical tests with which the user is familiar is bound to have a feedback effect at the stage at which the statistical model is shaped, as is suggested by the loop 2→3→4→2 in Figure 10.1. However carefully the model is draughted initially, subsequent experience at the testing stage may lead to a need to alter the model, frequently superficially, sometimes in substance. Many initially proposed models are found subsequently to be unsuitable for testing, so that the skill and patience of users is much exercised during earlier attempts at application.

At stage 3, choosing or constructing an appropriate statistical test need not be difficult once some fundamental decisions have been made. Usually it is fairly clear whether a one- or two-sample, or one- or two-tail test is needed. Where a choice is available between parametric and non-parametric tests in a particular situation, then the user may be influenced by the general factor that the parametric tests are more powerful, make more-demanding assumptions and require more-refined measurement scales and sampling plans than the non-parametric equivalents. Thus the user may like to decide whether the status of the study is 'preliminary' or 'definitive' by adopting *either* a rapidly applied, non-parametric test based on a small sample with an unrefined scale of measurement and a simple method of specimen selection *or* a generally more demanding and more laboriously applied parametric test. The user should carefully assess the labour of specimen selection and measurement against the overall objectives of the exercise, because the practice of economy of time and effort may have important influence. Why measure the lengths of hundreds of pebbles, each to the nearest millimeter, when simple ordering of three dozen in increasing length will suffice?

Only at stage 4 does the user actually try to 'do' anything practical. Stages 2 and 3 are planning stages which may not yet be free of defect. The user, however experienced, will sooner or later meet problems at stage 4. Whatever the simplicity or complexity of the situation in hand, when the user comes to apply the chosen test to the proposed model, defects of the logic of the model or some incompatibility of test with model may well be discovered. Stage 4 is a good point at which to check that all is well by verifying that the thinking behind the proposed approach is sound, that the sampling procedure is adequate, that the projected sample size is large enough or not too large, and that scales of measurement to be applied to the sample meet the requirements built into the tests.

It is likely that in practice the user has crystallised certain ideas arising from the stage 1 preliminary description, and has selected a test procedure that is formally or theoretically suited to his model, only to find that there is some reason why the experiment cannot be put into effect. For example, it may be that the sampling scheme cannot be applied because of difficulties of obtaining specimens, or because specimens obtained do not constitute a sample size that is sufficient or do not fit the predicted frequency distribution, etc. Only when the user is confident that the application is going to be workable should he embark on the sampling and measuring procedures. Common reasons for negative answers to the question posed at stage 4 (Fig. 10.1) include undue complexity of the null and alternative hypotheses, measurement scales and sample sizes failing to meet the requirements of the test, and assumptions about the populations that may be unwarranted. The difficulties are usually remedied by returning to stage 2 and either re-defining the statistical model or simplifying the hypotheses, or some combination of the two.

Having effected the sampling plan and applied the test, the user arrives at stage 5, where the most troublesome problem likely to arise is the discovery of an observed value of the test statistic close to a critical value. Of course, having defined a critical region by step 5 of the hypothesis-testing procedure of Section 2.3.5, the objective answer is simple: the observed value of the test statistic either *does* or *does not* fall into the critical region so that the null hypothesis is correspondingly either *rejected* or *accepted*. However, even the professional statistician is likely to be unhappy with such an uncertain situation, since the exact position of the boundary of the critical region is chosen by the user in the first place, and their only guide is their own assessment of the risks of being wrong (see the discussion in Section 2.4). Consequently, if the user is worried about this uncertainty, he is likely to be dissatisfied with the outcome of the testing procedure. Should this be the case, the remedy in theory is simple: the power of the test must be increased. There are two ways in which this might be accomplished. The first is to increase the sample size N, the second is to choose an intrinsically more powerful test. As discussed in Section 2.4, the power of a given test increases with the sample size thus decreasing the probabilities of type I and type II errors. If increase in the sample size is impractical, then recourse to a more powerful test may be advised. There are the attendant disadvantages, however, that such tests require more rigorous sampling plans and generally more-refined scales of measurement. In either case, a return to stage 2 is implied (Fig. 10.1).

The user satisfied with the outcome of applying the test arrives at stage 6. He is happy to accept either the null or the alternative hypothesis in the knowledge that the probability of its being wrong can be quantified. In short, he can associate a stated probability with the truth of his stated hypothesis. Together with a summary of his reconaissance, a description of his statistical model and sampling plan, and a reference to his chosen statistical test, this stated probability and hypothesis may then represent a concrete advance upon which he can report with complete objectivity and make subsequent use of in developing his overall interpretation.

10.2 Sampling: introductory, and a cautionary tale

To make a gross understatement, all that remains is to add some further comment on sampling problems, although I have left this most contentious problem till last. No matter to what we apply the statistical methods, the sampling plan is a fundamental part of the study. Previous illustration (Section 6.7) showed that the use of a clumsy sampling procedure can lead to bias in the selection of specimens, with consequent distortion of the outcome of the statistical tests. Such distortion might be totally hidden from the user and the readers of their report. As was stated, this is particularly true amongst the more-powerful, parametric statistical tests. It is essential in all applications, especially so amongst those that use the powerful tests, that *the sample should be as free as possible from bias*.

First though, a brief cautionary tale to show how insidiously sampling problems can arise. Returning from an excursion east of Edinburgh in February 1976, during a fuel crisis, a party of geologists wishing to maximise their investment in fuel decided to stop at Aberlady Bay to collect teaching material, despite sub-zero temperatures and total darkness. To sample pebbles, the field car was driven on to the beach and two members of the group ('SC' and 'KRG') were issued with sacks. From the patch of beach illuminated by the vehicle lights, they were instructed verbally to collect 'about 100' pebbles each 'at random'. On returning to the vehicle, they complained bitterly of the harsh conditions and reported many pebbles firmly frozen in place.

Back in the laboratory, the 'length' (i.e. maximum dimension) of all specimens in each of the two samples was measured and the results presented graphically in the form of cumulative distribution functions (Fig. 10.2). The fact that the sack of 'KRG' pebbles was considerably heavier than that of the 'SG' pebbles immediately suggested sampling bias and the displacement of the 'KRG' curve to the right with respect to the 'SC' curve in Figure 10.2 might support a proposal that the 'KRG' pebbles are, in general, larger. In fact, using the latter as a basis of an alternative hypothesis, the Kolmogorov–Smirnov two-sample, one-tail test was applied, leading to rejection of the null hypothesis with a critical region of

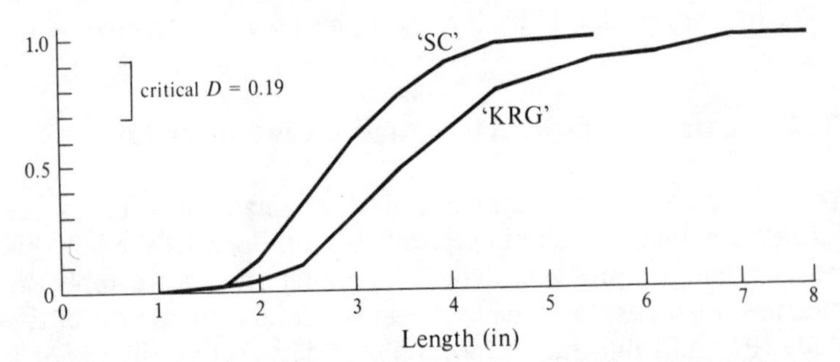

Figure 10.2 Cumulative distribution functions for the variable 'length' for two samples of pebbles hand picked from the same patch of beach. See text for discussion. The length of the vertical bar shows the critical value of the Kolmogorov–Smirnov two-sample test statistic D.

size 0.01. Yet despite this rejection, eight witnesses in the field car saw the two collectors sample exactly the same area of beach. The conclusion arises that the simple verbal instruction to collect 'at random' is inadequate, and that severe sampling bias has been introduced in this process of selective hand picking. In this instance, the bias might partly be accounted for in the difference in stature of the collectors: 'SC' is a slight 5 ft 4 in, 'KRG' a well built 6 ft 7 in with corresponding differences in hand sizes.

Another interesting feature of the two samples is that, although they are almost certainly biassed strongly with respect to 'length', both can be compared separately with Gaussian distributions. On application of a chi-squared test, a null hypothesis to such effect can be accepted with a critical region of size 0.01, so that the two samples appear, separately, to have acceptable Gaussian distributions. Consequently, parametric tests such as Student's t for the difference of two means can be applied, leading to rejection of a null hypothesis of equal means with a critical region of size 0.001.

10.3 Sampling: possible solutions

Practical statisticians have devised a number of sampling schemes, several marked by great ingenuity, and all with specific objectives. The geologist may use these schemes, but it may be difficult because geological materials and observations are frequently hard won under difficult circumstances. Indeed, by comparison, the statistician practising in the more homely fields of biometrics or the social sciences usually has less difficulty in choosing and increasing sample size by seeking further cases, growing more potatoes, etc. The geologist, contrarywise, often has to wrest his specimens or observations from a well consolidated matrix under difficult, even hazardous physical situations and samples are often both small and unrepresentative. At best, the geologist must be alert to possible sources of bias and try to devise some way of reducing, even eliminating the effect. In practice, the variety of situations in which bias might arise in sampling geological materials are so numerous that it would be meaningless to attempt a listing here: acute observation and agile wit on the part of the collector are useful first lines of defence.

10.3.1 Reasons for drawing samples

A sample may be drawn from a population purely for descriptive purposes, the specimens being measured and those measurements presented in tabular or graphical form, with no further processing, as being descriptive and representative of their parent population. Any bias in the sample leads to a biassed impression of the population.

Every population may be characterised by a parameter, a ratio, a proportion or some other number whose value is estimated from measurements on the sample. Since such estimates are generally made from a sample that may be a small part of the population, then the critical observer will ensure unbiassed sampling and will provide some guidance as to the precision of the estimated parameter, ratio, etc., via a stated confidence interval. A large part of sampling theory is concerned with estimators and their precision. We have had some discussion under the heading of 'confidence intervals'.

Lastly, we have devoted much time to hypothesis testing. The methods are based on samples and defective sampling practices can lead to erroneous decisions. Generally, therefore, sampling may have either descriptive or analytical objectives.

10.3.2 Populations: target and sampled, infinite and finite

An essential early step in a sampling plan is to define, sometimes rigorously, the characteristics that a specimen must display in order to be included within the population of interest. This may be difficult, as when sampling 'granite' in an area in which development of migmatites has introduced a variety of transitional rock types, some of which may not strictly be 'granite'. Elsewhere, the task may be straightforward, as when sampling dykes intruded into sedimentary rocks. Additionally, the observer has to consider carefully whether the whole of their defined population is, in fact, available to be sampled. Cochran (1977, p.5) uses the terms 'target' and 'sampled' populations. Ideally, these two populations coincide but in practice this may be difficult or impossible to arrange. Thus, rock formations may be buried deeply, even unconformably overlain by younger sedimentary deposits. Although technically accessible, as by diamond drilling, costs may dictate limits to the sampled population. Similarly, parts of formations now eroded are inaccessible. Consequently, an observer in such a position must give careful consideration as to the amount of overlap of their 'target' and 'sampled' population and assess problems arising if this overlap is small.

Until now, we have considered 'populations' as consisting of an infinite number of specimens or individuals. Sometimes this may not be reasonable as, for example, the population of sandstone beds in the Carboniferous rocks of central Scotland: there is a finite number of such beds. Thus, we recognise 'infinite' and 'finite' populations. In practice, the boundary between the two is arbitrary. So long as our sample constitutes a fraction no greater than one twentieth of the finite population, then we can safely take the latter as infinite. Difficulties might arise with certain sampling schemes and certain objectives but they lie outside the scope of this book. They are likely to be of importance to researchers or geologists whose brief is highly specific and who will find Cochran (1977) useful reading.

10.3.3 Principal considerations in devising a sampling scheme

Many sampling schemes have been devised in which an essential element of pure chance enters into the selection of individuals or specimens. Such schemes constitute methods of 'probability' sampling which are to be contrasted with 'non-probability' methods, such as hand-picking or the subjective selection of 'typical' specimens. These last two cannot be used for any objective assessments of populations.

Having defined the overall objectives of the enterprise and stated clearly the boundaries of the population to be sampled, the frame of reference which will be used to sample the population must be decided. In many cases, specimens will be selected so that they cover the population more or less uniformly: granite specimens will cover the granite outcrop with uniform density, gabbro specimens will be taken from the layered sequence so that they cover the stratigraphical height with uniform density. However, in other cases, specimens will not cover an area or length uniformly, even though they cover the population uniformly. This will be the case, for instance, if the variable of interest is connected with the distance from the margin of a circular outcrop or is connected with stratigraphical height in a sequence that has been folded recumbently. Consequently, the population has to be thought of as being 'staked out' or otherwise covered with a reference frame. Formally, this operation consists of dividing the population into a series of mutually exclusive units, or areas, or increments, all of which may differ in physical dimensions from one part to another. Elsewhere, the operation may consist of constructing reference axes along whose direction the units of measurement may differ. The discussion of some examples below (Section 10.3.4) will illustrate this seemingly obscure point.

A final, but important step is to assess the degree of precision required (in a descriptive application where parameters are estimated) or to assess the size of the critical region (in an analytical application where hypotheses are to be tested). Factors to consider will include the scale of measurement to use on the variable of interest and the overall effort or cost that is realistic to the objectives. Decisions such as these are helped considerably by previous experience, by advice or the running of a 'preliminary test' or 'pilot study' in advance of the main exercise.

10.3.4 Some specific sampling schemes

As we noted, pure chance is an essential ingredient of any specimen selection procedure, but we also noted that a variety of schemes or procedures have been developed. The principal motivation for the diversity lies in a need to achieve maximum efficiency and precision within constraints such as cost, time, effort, available facilities, manpower skills, specimen accessibility and so on. The result is that the researcher may be faced with a bewildering choice. However, with clear recollection that the objective of the methods we develop here are that they be applied 'on site' and with minimal practice of sophistry, there is a relatively small number of sampling schemes that may be introduced with brevity.

Simple random sampling. Most 'on-site' applications will adopt this method, and almost certainly all of those intended for hypothesis testing. The method is thought of as attaching a unique identifying number to each specimen in the population and then drawing a sample of the required size by generating a list of uniformly distributed random numbers (i.e. numbers which have equal probabilities of occurrence over the range of interest). For example, suppose we wish to draw a sample of 30 pebbles from a beach or river gravel surface. One method would be to stretch out in a straight line a measuring tape, say 36 yards long (the reason for Imperial rather than metric measure lies in the ease with which this interval can be divided by 6, several times over). To generate random numbers, suitable tables could be used and are abundantly available in standard texts (e.g. Yule & Kendall, 1950 and 1.25 kg; Meyer, 1975 and 1.16 kg). However, many of my target readers will be practical people, carrying poker dice (20 g) for those periods of (geological) inactivity that characterise any field project. Each face of a die can be identified with a number in the range 1 to 6 inclusive, and casting the die will produce one of these numbers at random.

This random number can be identified with one of the six 6-yard intervals into which our 36-yard tape can be divided. Casting a second die will identify a 1-yard interval within the specified 6-yard interval and so on, dividing each lesser interval into six parts to whatever ultimate interval is needed. The pebble that occupies this interval is then selected to join the sample. Thus the dice sequence 1, 3, 6, 4 will identify a pebble 2 yards and 33 inches along the tape. A simple random sample of size 30 drawn by this method is shown in Figure 10.3.

One problem that may arise from time to time is that the same specimen becomes selected more than once. Thus we identify sampling *with* replacement and sampling *without* replacement depending on whether or not we replace that specimen as we continue sampling. It may appear odd to have the same specimen represented more than once in a sample but the statistical reasons are sound. In practice, so long as the sample is less than about one twentieth of a finite target population, then the difference is academic.

Stratified random sampling. Some observers may feel uneasy with the way in which simple random sampling may tend to cluster specimens over part of the sampled range and leave large gaps elsewhere, as well as duplicating specimens on occasion if sampling is with replacement. A method of reducing this effect is to divide the population into a number of mutually exclusive 'subpopulations', or 'strata', over the range of interest and to apply the simple random sampling technique to each stratum in turn. In our hypothetical example of sampling beach pebbles over a 36-yard

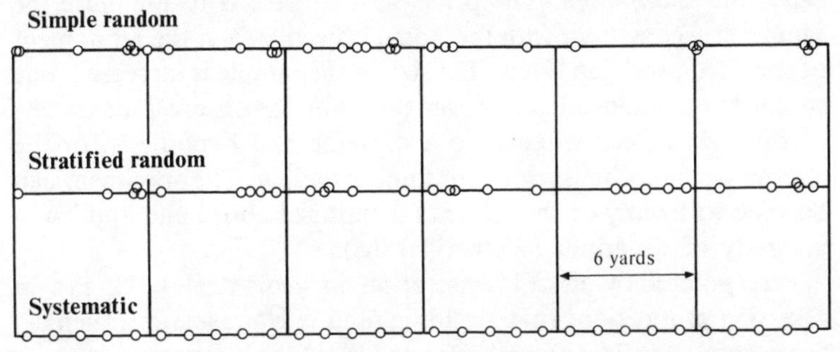

Figure 10.3 Thirty sampling points along a 36-yard interval drawn by a simple random sampling scheme, a stratified random sampling scheme (in which each stratum is 6 yards long) and a systematic sampling scheme.

interval, Figure 10.3 shows how specimens are distributed, five to each 6-yard stratum.

However, there is often a more fundamental reason for adopting a stratified random sampling scheme, namely to improve on the precision of estimation of some population parameter. A graphic example is provided by the data used in Figure 9.9. The aim was to estimate the axial direction of the fold together with its cone of confidence. To this end the sampling was arranged to provide an approximately uniform distribution of points along the girdle, i.e. the population was stratified according to orientation. The result gave a cone of confidence of near circular form, although the fold in question has long, nearly straight limbs and a restricted hinge area. Simple random sampling of such a structure would have provided a distinct bimodal distribution of points on the girdle, coresponding to the fold limbs, and an estimate of the axial direction with the cone of confidence distinctly elliptical in form. Similar examples concerned with estimating means, etc. are discussed by Cochran (1977).

Systematic sampling. As with stratified sampling, this method requires the population be stratified, but now with as many strata as there are to be specimens in the sample. The position of the first specimen, taken from the first stratum, is decided at random, but then all succeeding specimens are taken from the same position in their respective strata. Figure 10.3 illustrates the arrangement.

Systematic sampling is attractive in reducing the amount of labour involved, but systematic samples provide estimates of generally lower precision than do the other methods.

Sequential sampling. The principal aim here is to minimise the sample size consistent with the stated objective, e.g. the attainment of specified precision levels. The size of the sample is increased, one specimen at a time, and after every addition the observer makes one of three decisions: whether to accept the null hypothesis, or the alternative hypothesis, or to continue sampling. The specimens can be selected by any of the schemes mentioned above and applied in a variety of situations (Wetherill 1966).

One potentially useful application in geological work lies in assessing proportions that are to be used in classification, such as, for example, deciding whether a rock in thin section contains more or less than 20% of quartz by volume and is thus to be classed as 'granite' or 'quartz syenite'. 'On site', it is possible to use the

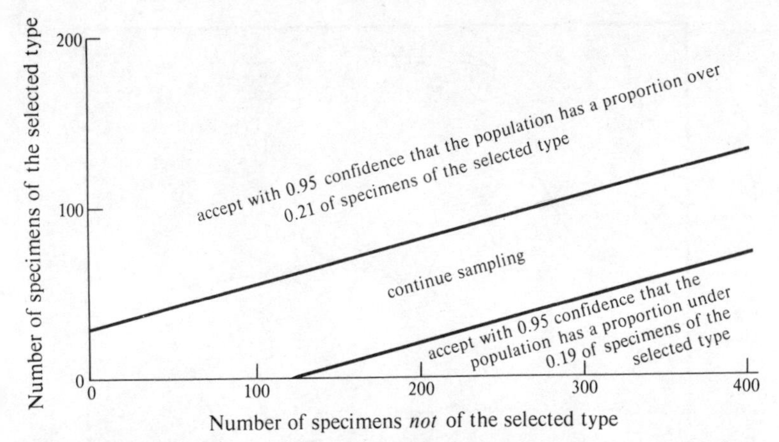

Figure 10.4 A chart for the sequential testing of a proportion of 20%. Before setting out to use such a chart, the observer has to accept that it is impossible to decide whether a population has a proportion of selected specimens that is greater than, or less than 20% *exactly*. In practice, they have to accept that the true proportion is greater than say, 21% or less than say, 19%, or to continue sampling. The chart is used thus: starting at the origin, the user marks a series of straight-line increments parallel to the axes, one for each specimen drawn. If the specimen is of the selected type (e.g. point on 'quartz' in a point count), then the increment is marked vertically upwards. If the specimen is not of the selected type (e.g. point not on 'quartz'), then the increment is marked to the right. Once the 'tip' of the stepped graph so constructed emerges from the zone of continued sampling, one or other of the two hypotheses may be accepted. The general form of the straight-line components of the chart is shown. Clearly, one chart can carry more than one pair of such lines and thus be used in a many-way classification exercise.

The general equation of the lines is

$$y = \pm\frac{\log((1-a)/a)}{\log(p_1/p_0)} + \left\{\frac{p}{1-p}\right\}x$$

with $1 - a =$ confidence interval, p_1, $p_0 =$ upper and lower bounding proportions, and $p =$ proportion under test.

scheme to decide, for instance, whether a particular ore mineral is sufficiently concentrated for economic exploitation given that proportions can be estimated by the 'point-counting' method. The scheme is simple if a chart such as that illustrated in Figure 10.4 is constructed.

10.3.5 An example: some sampling problems at St Baldred's Cradle

St Baldred's Cradle (longitude W2°35′, latitude N56°01′) is a low headland lying on a wavecut platform on the southern coast of the Firth of Forth, south-east Scotland (Fig. 10.5). A near circular out-

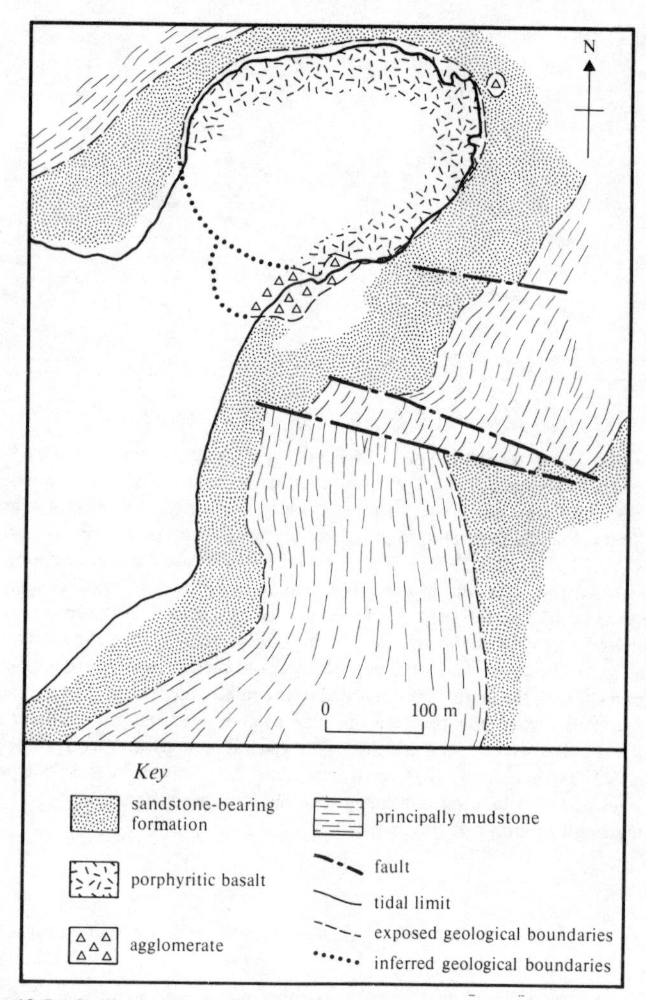

Figure 10.5 Outline geological map of the vicinity of St Baldred's Cradle, East Lothian, Scotland. Areas left blank are not exposed.

crop of Carboniferous porphyritic basalt 200 m in diameter, is surrounded by Lower Carboniferous sedimentary rocks, mostly clastic, which have low dip, except near the basalt where they dip concentrically inwards at angles up to 40°. Agglomerate is present locally around the margins of the basalt outcrop. A number of interesting sampling problems arise in detailed study of the area.

First, the basalt: although having a fine-grained matrix, this rock contains some components whose concentrations are quantified easily in the field, namely: vesicles and phenocrysts of augite and olivine. Some of the last are replaced by iddingsite. On the basis of subjective visual comparison, the proportions of all of these components seem to differ from place to place over the basalt outcrop. Given a geological model that the proportions of phenocrysts etc. are a function of distance from the margin of the outcrop, which seems at least partly intrusive, then the sampling scheme has to take note of the circular form. Assuming a scheme of systematic sampling, then Figure 10.6 shows that the sampling frame has to be constructed carefully if the central parts of the outcrop are not to be over-represented volumetrically. Thus, with specimens collected along a diametrical traverse across the outcrop, even the systematic scheme will specify sampling points at non-regular intervals, illustrating the importance of defining the sampling frame as mentioned in Section 10.3.3. If increments of equal area are to be sampled uniformly, then the distance of the specimen locations from the centre of the outcrop are to be *non-uniformly* distributed. Figure 10.7 shows how these distances may be generated randomly if a scheme of simple random sampling is specified. The general method of drawing specimens at random from a non-uniform distribution is given by Meyer (1975, pp.167–9), of which this is an adaptation. A similar method may be used to generate random selections of numbers from a Gaussian or any other population, should these be needed.

Still with the basalt, the volume proportions of vesicles or of the two phenocryst phases are estimated readily by a point-counting

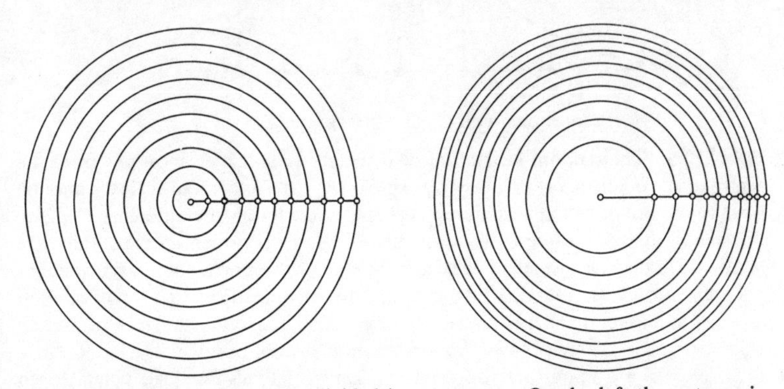

Figure 10.6 A circular outcrop divided into ten parts. On the left the outcrop is divided by circles whose radii have a constant difference; on the right, by circles drawn such that the annuli so defined are of equal area. A sample traverse across a diameter of the outcrop must have specimen localities tied to the grid on the right if it is to be volumetrically unbiassed with respect to distance from the margin of the outcrop.

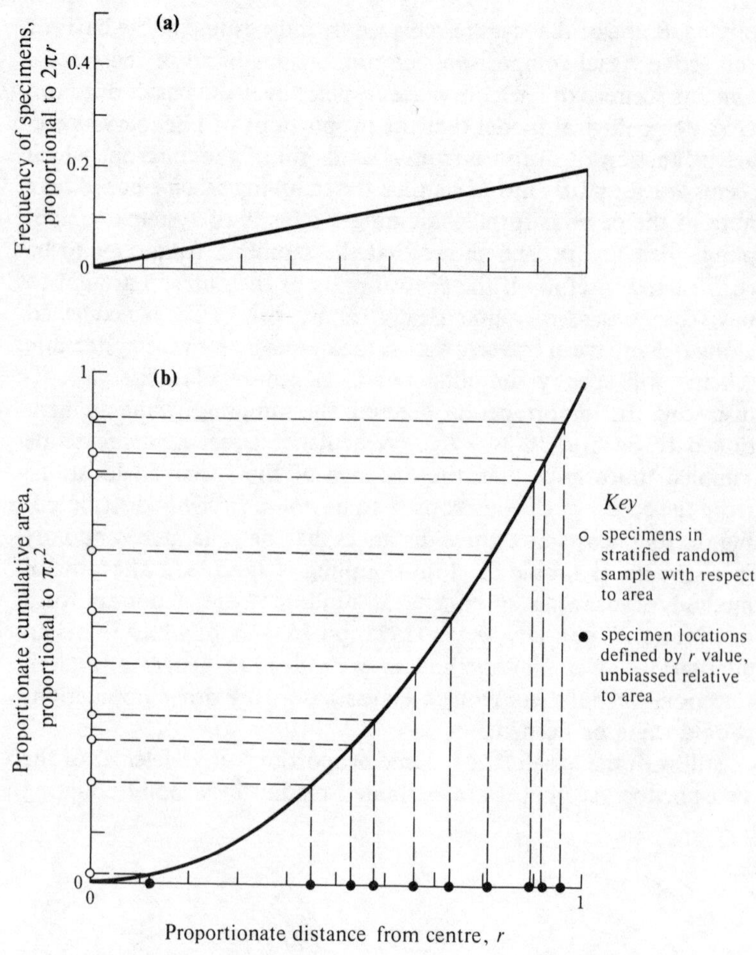

Figure 10.7 The random generation of distances of specimen locations from the centre of the outcrop. (a) Frequency of specimens as a function of their distance (r) from the centre of a circular outcrop is proportional to the circumference $C = 2\pi r$. For small values of r, there are few specimens because the circumference of the circle along which they lie is small. The largest number of specimens for a given value of r is found around the margin of the outcrop. (b) Cumulative distribution of outcrop area (A) as a function of r, obtained by integrating the function above. Any random value of fractional area drawn from a uniform distribution will correspond to a value of r in a non-uniform distribution. Thus randomly sampled points (open circles) uniformly distributed in area can be located on the ground according to their distance from the outcrop centre (closed circles).

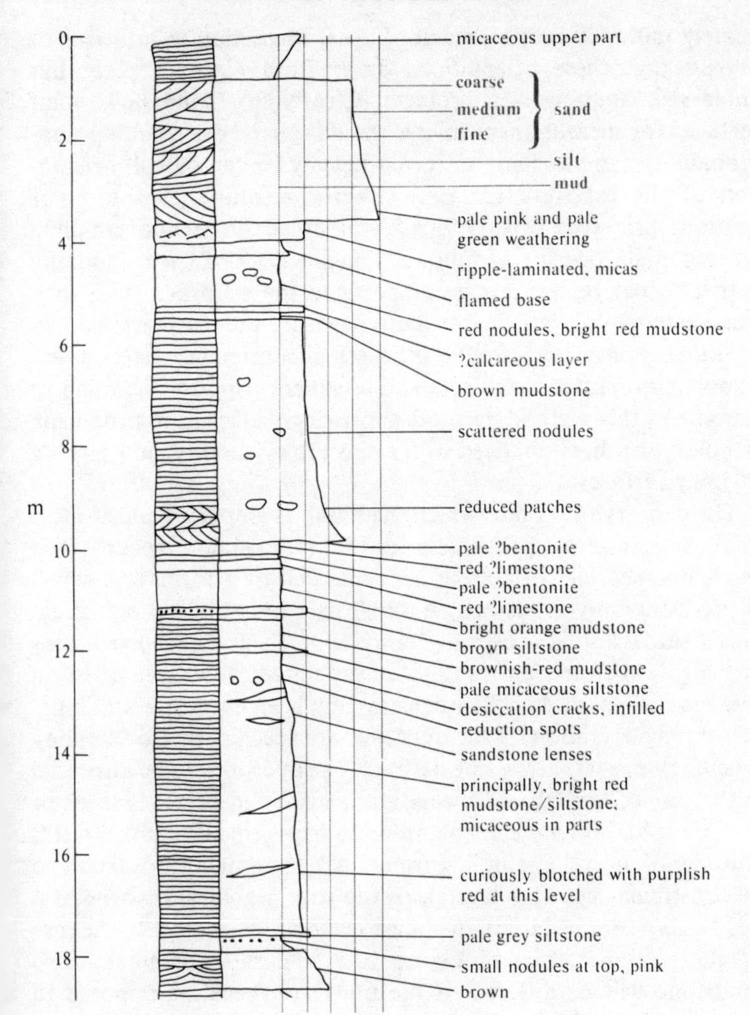

Figure 10.8 Details of a small part of the sedimentary succession at St Baldred's Cradle, East Lothian, Scotland.

method and confidence intervals approximated (Section 5.9). If these proportions are to be used in any classification scheme, e.g. under 5 % of olivine phenocrysts replaced by iddingsite classed as 'slightly altered', over 5 % classed as 'moderately altered', then a sequential sampling chart may be drawn up.

Finally, exposures of the basalt are characterised by prominent joint surfaces. The joints seem to occur as sets of subparallel fractures, each exposure typically containing three such sets, approximately mutually perpendicular. Casual inspection is sufficient to suggest that these orientations differ from place to place but unbiassed sampling is a problem. Clearly, to 'hand pick' joint surfaces for measurement invites the danger of the resulting non-probability sample being distorted heavily by the overall orientation of the exposure surface, with the resulting generation of spurious preferred orientations. A modified systematic sampling scheme might operate as follows. Imagine a set of fixed, mutually perpendicular reference axes adjacent to the exposure, with each axis being of unit length (say 1 m). Measure the orientations only of those joints that would intersect the reference axes if extrapolated. With a meter rule and a camera tripod with which to support it, this method becomes surprisingly efficient in producing samples tolerably unbiassed with respect to orientation and spacing of joint surfaces.

The country rocks into which the basalt is emplaced constitute a thick sequence of principally clastic sedimentary rocks. Two sandstone-bearing formations are separated by a formation which is predominantly mudstone. A small part near the top of the exposed succession is shown in Figure 10.8. Such detailed sedimentary logs will clearly answer almost any statistically based question that may be posed, but their construction is an expensive and time-consuming operation. If the questions are specific, then a sampling scheme can nearly always be devised. The questions to be answered in the case of any layered succession, be it sedimentary, igneous or metamorphic will have in common the requirement to give careful consideration to the sampling frame. Are the units of this frame to be constituent beds (or layers), or the stratigraphical or structural height above some datum, or some system of facies etc.? In the case of the section shown in Figure 10.8, the ratio of siltstone to mudstone will be different if the units are 'beds' as opposed to 'stratigraphical thicknesses'.

10.3.6 Concluding remarks

Sampling geological materials can be difficult, even in an attempt to answer elementary questions. Such is the rich diversity of rocks, minerals, fossils and structures that it would be an impossible task to give guidelines on any general approach. Krumbein and Graybill (1965) and Koch and Link (1970) provide some interesting discussion and give examples of specific schemes, but unless a researcher or serious investigator is fortunate in discovering a ready-prepared method, recourse may well have to be made to first principles in some elementary text such as that of Cochran (1977).

Appendix A Matrix and vector algebra – a glossary

Extensive cross referencing may be needed in the use of this glossary. For more explicit accounts, see either Frazer, Duncan and Collar (1963) or Hall (1963). The latter is an excellent handbook but many of the more modern statistics texts include a chapter on the subject.

addition Matrices to be added must be of the same **order**. If the sum of two matrices **A** and **B** is **C**, then: $\mathbf{C} = \mathbf{A} + \mathbf{B} = \mathbf{B} + \mathbf{A}$. The **elements** of **A**, **B** and **C** are related by $C_{ij} = A_{ij} + B_{ij}$.

bordering A procedure by which additional **rows** or **columns** or both are appended to a **matrix** so that its **order** is increased.

column, column matrix A single column of **elements** in a **matrix**. A **matrix** of **order** (P by 1) where P is an integer.

component An **element** of a **vector**.

determinant A function of a **matrix** expressable as a single number. For the determinant of a **square matrix** of **order** (3 by 3), proceed as follows. Write down the matrix and repeat its first two columns to the right:

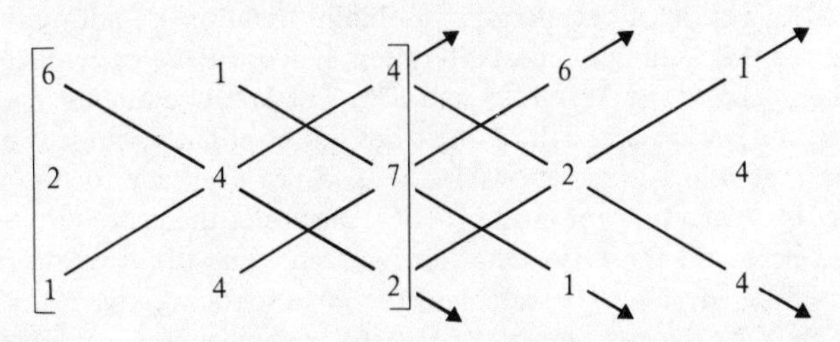

Multiply together the three numbers along each of the three descending arrows to obtain: 48, 7 and 32. Add these three: 87. Repeat for the ascending arrows, obtaining 16, 168 and 4, adding up to 188. Subtract the figure for the ascending arrows from the figure for the descending arrows to obtain – 101, the required determinant. For the determinant of a square matrix of order (2 by 2), proceed in much the same way.

diagonal Any set of **elements** falling from top left to bottom right of a **matrix**, such as T_{11}, T_{22}, T_{33}, etc. or any set parallel to it.

eigenvalues, eigenvectors Associated with a **square matrix** of **order** (P by P) are P eigenvalues, with each of which, in turn, is associated an eigenvector of P **components**. In the case where the **matrix** is **symmetrical** and all the **elements** of the **principal diagonal** are positive, then, if the matrix is of order (3 by 3), it can be represented graphically as an ellipsoid. The eigenvectors then correspond to the principal axes of the ellipsoid and the eigenvalues correspond to the lengths of the principal axes. Calculation of eigenvalues and eigenvectors is illustrated in Table A.1 referring to a matrix of order (3 by 3) with co-ordinate reference axes pointing north, east and down. Angles made by **vectors** with these axes are denoted by X_n, X_e and X_d respectively and are measured from the positive ends of the axes. The eigenvector associated with the largest eigenvalue is calculated first (Table A.1a). The first entry to make in the tabulation is the matrix **T**. Next, on the left, enter the angles made by an approximation to the required eigenvector with the 'north', 'east' and 'down' axes. It does not matter how crude this approximation is, the method will always converge ultimately on the required solution. However, the better the approximation, the shorter the task. Next, enter the cosines of these angles in the table. They are +0.809, −0.588 and +0.087 and are the **components** of the **vector** that has been taken as the approximate solution. Now **pre-multiply T** by the **row matrix** consisting of the components of the approximate solution to obtain the product: +5.447, −4.086, +0.490. These three numbers are the components of a new vector. **Normalise** these components and enter them in the next row on the left. They are +0.798, −0.598 and +0.072 and, because they have been normalised, they are the direction cosines of a vector that is slightly nearer to the required eigenvector than our first approximation. To complete 'iteration 1', look up the angles that correspond to these direction cosines and enter then in the table: 37°, 127° and 86°. Repeat this cycle of calculations to obtain 'iteration 2', noting that the value for the X_d angle repeats iteration 1. Repeat the cycle once more to obtain the angles 38°, 128° and 87°. The first two angles repeat the values of iteration 2, the third angle has moved slightly, but was preceded by repetitions in iterations 1 and 2. Since accuracy to the nearest degree is satisfactory for most practical purposes, the procedure is terminated here and the third iteration is accepted as the solution for the required eigenvector. To calculate the corresponding eigenvalue, pre-multiply **T** by the components of the accepted eigenvector to obtain the product: +5.313, −4.245, +0.349. The required eigenvalue is the length of this vector, calculated as the square root of the sum of the squares of the three components. Thus, the maximum eigenvalue, τ_3, is: $\sqrt{(5.313^2 + 4.245^2 + 0.349^2)} = 6.810$.

To calculate the eigenvector corresponding to the minimum eigenvalue, first calculate the matrix $\mathbf{T} - \mathrm{tr}(\mathbf{T}) \times \mathbf{I}$ where $\mathrm{tr}(\mathbf{T})$ is the **trace** of **T** and **I** is the **unit matrix** of order (3 by 3). Enter this matrix as in Table A.1b. The calculation proceeds exactly as before and is terminated when iterations repeat themselves. There are two important differences, however. First, the vectors generated by pre-multiplying the matrix at the top of Table A.1b by the approximate solutions have their third component negative, e.g. −9.389. In normalising these vectors, first multiply through their components by −1 so that the third component of the normalised vector is always positive and X_d is thus always in the lower hemisphere of the stereographic projection. The second difference

Table A.1 Stages in the calculation of eigenvalues and eigenvectors.

	X_n	$\cos X_n$	X_e	$\cos X_e$	X_d	$\cos X_d$			
(a)						matrix $\mathbf{T}$ =	+5.378	−1.616	+1.680
							−1.616	+4.980	+1.721
							+1.680	+1.721	+1.641
approximate solution	36	+0.809	126	−0.588	85	+0.087	+5.447	−4.086	+0.490
iteration 1	37	+0.798	127	−0.598	86	+0.072	+5.379	−4.144	+0.430
iteration 2	38	+0.791	128	−0.609	86	+0.063	+5.344	−4.203	+0.384
accepted solution	38	+0.785	128	−0.617	87	+0.056	+5.313	−4.245	+0.349
(b)						$\mathbf{T} - \mathrm{tr}(\mathbf{T}) \times \mathbf{I}$ =	−6.622	−1.616	+1.680
							−1.616	−7.020	+1.721
							+1.680	+1.721	−10.359
	X_n	$\cos X_n$	X_e	$\cos X_e$	X_d	$\cos X_d$			
approximate solution	112	−0.375	123	−0.545	41	+0.755	+4.632	+5.731	−9.389
iteration 1	113	−0.388	119	−0.480	38	+0.787	+4.667	+5.351	−9.630
iteration 2	113	−0.390	117	−0.447	36	+0.805	+4.657	+5.154	−9.763
iteration 3	113	−0.388	115	−0.430	36	+0.814	+4.632	+5.047	−9.824
accepted solution	113	−0.387	115	−0.421	35	+0.820	−0.023	−0.060	−0.029

lies in the calculation of the minimum eigenvalue τ_1. Once an accepted solution for the eigenvector has been obtained, the original matrix **T** must be pre-multiplied by the vector components to obtain the last entries in Table A.1b: -0.023, -0.060, -0.029. The minimum eigenvalue is then the square root of the sum of the squares of these three numbers: $\tau_1 = 0.0705$.

The eigenvector corresponding to the intermediate eigenvalue is perpendicular to the two eigenvectors already calculated (which are themselves perpendicular, we hope!). It can thus be obtained by simple graphical construction on the stereographic projection and is found to be: $X_n = 61°$, $X_e = 48°$ and $X_d = 55°$, with corresponding direction cosines (or components): $+0.482$, $+0.665$, $+0.569$. The intermediate eigenvalue is then obtained by pre-multiplying **T** by these vector components to obtain: $+2.473$, $+3.512$ and $+2.888$ and then taking the square root of the sum of the squares of these three numbers, as before: $\tau_2 = 5.176$.

As a final check on the arithmetic:

$$\tau_1 + \tau_2 + \tau_3 = \mathrm{tr}\,(\mathbf{T})$$

element This refers to any of the entries that go to make up a **matrix**. A convenient form of labelling elements of a matrix is to give the number of the **row** and the number of the **column** as suffixes. Thus, T_{rc} is the element in the rth row and the cth column of the matrix **T**. The usual convention is that the first suffix is the row number and the second suffix is the column number. Thus, T_{12} is in row 1 and column 2 and is read as '*T* one-two'.

inverse See **reciprocal**.

matrix An array of quantities arranged in tabular form and vested with a variety of properties. The elements may be unknown or may be functions as well as numerals:

$$\text{e.g.}\quad \mathbf{T} = \begin{bmatrix} \sin y & 0 & -1 \\ 10 & e^x & 7 \end{bmatrix}$$

In practice, most of our matrices are **square**.

minor, principal minor A minor is a **matrix** derived from a pre-existing matrix by omitting one or more row – column pairs. If the row – column pair contains an element from the **principal diagonal** of the parent matrix, the minor is a principal minor.

multiplication Matrices may be multiplied by constants or by other matrices. When multiplied by a constant, every **element** of the matrix is multiplied by the constant. Thus, 12 times the **unit matrix** of **order** (3 by 3) is:

$$12 \times \mathbf{I} = 12 \times \begin{bmatrix} 1 & 0 & 0 \\ 0 & 1 & 0 \\ 0 & 0 & 1 \end{bmatrix} = \begin{bmatrix} 12 & 0 & 0 \\ 0 & 12 & 0 \\ 0 & 0 & 12 \end{bmatrix}$$

There are two important rules in the multiplication of a matrix by another matrix. First, there is a relationship between the **orders** of the two matrices and their product such that for the equality $\mathbf{C} = \mathbf{A} \times \mathbf{B}$, if $\mathbf{A}$ is of order (P by Q) then $\mathbf{B}$ must be of order (Q by R) and the product $\mathbf{C}$ is then of order (P by R). The second rule governs the evaluation of the elements of the product and is first illustrated in the case of $P = R = 1$; $Q = 2$, i.e. a row matrix is multiplied by a column matrix:

$$\mathbf{B} = \begin{bmatrix} 3 \\ 4 \end{bmatrix}$$

$$\mathbf{A} = \begin{bmatrix} 1 & 2 \end{bmatrix} \begin{bmatrix} 11 \end{bmatrix} = \mathbf{C}$$

In this simple example, $\mathbf{A}$ is multiplied by $\mathbf{B}$ to give a product $\mathbf{C}$ of order (1 by 1), i.e. a single number. The rule for the evaluation of this number is: $(1 \times 3) + (2 \times 4) = 11$, i.e. the first elements of $\mathbf{A}$ and $\mathbf{B}$ are multiplied, the second elements are multiplied and the two products obtained are added together.

This simple rule of multiplication is extended to matrices of higher orders thus:

$$\mathbf{B} = \begin{bmatrix} 3 & 2 \\ 4 & 0 \end{bmatrix}$$

$$\mathbf{A} = \begin{bmatrix} 1 & 2 \\ 0 & -1 \end{bmatrix} \begin{bmatrix} 11 & 2 \\ -4 & 0 \end{bmatrix} = \mathbf{C}$$

Elements in $\mathbf{C}$ are obtained by multiplying the corresponding row in $\mathbf{A}$ by the corresponding column in $\mathbf{B}$. In the case of **square matrices**, note that the sequence of multiplication is important. In this example, $\mathbf{B}$ has been pre-multiplied by $\mathbf{A}$ (or conversely, $\mathbf{A}$ has been post-multiplied by

B). If the sequence is reversed, then the product is generally different:

$$\mathbf{A} = \begin{bmatrix} 1 & 2 \\ 0 & -1 \end{bmatrix}$$

$$\mathbf{B} = \begin{bmatrix} 3 & 2 \\ 4 & 0 \end{bmatrix} \begin{bmatrix} 3 & 4 \\ 4 & 8 \end{bmatrix} = \mathbf{D}$$

Thus, **A** × **B** is not equal to **B** × **A**.

normalisation (of a vector) The length of a **vector** is equal to the square root of the sum of the squares of its components. Thus, a vector with components 5.447, −4.086 and 0.490 has a length of 6.827. The vector is said to be normalised when the components are adjusted by dividing each of the components by the vector length, so that the normalised vector is of unit length.

order This is a means of expressing the size of a **matrix**. A matrix of order (P by Q) thus has P rows and Q columns.

partition In certain matrix operations, it is necessary to split off groups of rows or columns. This process is known as partitioning.

post-multiply, pre-multiply See **multiplication**.

principal diagonal The diagonal of elements running from the top left element of a matrix and thus comprising the elements T_{11}, T_{22}, T_{33}, etc.

reciprocal For every **matrix** that has a **determinant** not equal to zero, there is a corresponding reciprocal, or inverse, **matrix** such that the product of a matrix and its inverse is the **unit matrix**. Calculation of reciprocal matrices is involved; a simple method is given by Frazer, Duncan and Collar (1963).

row, row matrix A single row of **elements** in a **matrix**. A matrix of **order** (1 by Q) where Q is an integer.

square matrix A matrix in which the numbers of rows and columns are equal.

symmetrical matrix A matrix symmetrical about the **principal diagonal** such that the **element** T_{ij} is equal to the element T_{ji}. An example is:

$$\begin{bmatrix} 1 & 3 & 5 \\ 3 & 0 & -1 \\ 5 & -1 & 6 \end{bmatrix}$$

trace The sum of the **elements** in the **principal diagonal**. Thus for the matrix above, the trace equals $1 + 0 + 6 = 7$.

transpose The transpose, **T**′ of a matrix **T** is obtained by writing the columns of **T** as the rows of **T′**. Thus, if:

$$\mathbf{T} = \begin{bmatrix} 1 & 3 \\ 5 & 7 \end{bmatrix} \quad \text{then} \quad \mathbf{T}' = \begin{bmatrix} 1 & 5 \\ 3 & 7 \end{bmatrix}$$

unit matrix The matrix equivalent of the digit one. A **square matrix** in which the **elements** of the **principal diagonal** are unity and all other elements are zero. Symbolised by **I**. Thus the unit matrix of **order** (3 by 3) is:

$$\mathbf{I} = \begin{bmatrix} 1 & 0 & 0 \\ 0 & 1 & 0 \\ 0 & 0 & 1 \end{bmatrix}$$

vector A physical quantity in which magnitude is associated with direction, e.g. water current velocity. The **components** of a vector are its projections on to two or three co-ordinate axes (depending on whether the problem has two or three dimensions) as shown in Figure A.1. These components can be entered as **elements** in a **column matrix**, or more commonly, in a **row matrix**. A vector of length unity is a 'unit vector'.

Vector addition consists of putting the vectors head to tail without rotating them and calculating the vector that joins the tail of the first to the head of the last. This is the 'resultant' and its components R_n and R_e are the sums of the components of the constituent vectors $\mathbf{l}_1$ and $\mathbf{l}_2$:

$$R_n = l_1 \times \cos X_{n1} + l_2 \times \cos X_{n2}$$
$$R_e = l_1 \times \cos X_{e1} + l_2 \times \cos X_{e2}$$

where X_{n1} is the angle vector $\mathbf{l}_1$ makes with the 'north' axis, etc.

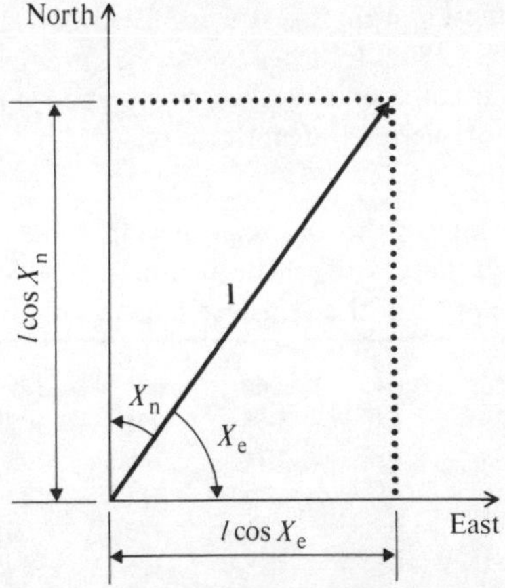

Figure A.1 The projection of a vector in two dimensions on to mutually perpendicular reference axes. The vector components are $l \times \cos X_n$ and $l \times \cos X_e$.

Appendix B Further exercises

In the earlier chapters especially, there have been a few exercises interspersed with the text and intended to consolidate understanding of certain of the methods developed. Here, we have a small collection of additional exercises, one or two of which expand those already met, but mostly they are of a comprehensive nature and span the contents of two or more of the chapters into which the book is necessarily artificially divided. These exercises are set out in a way in which they might be explored in reality and in which the reader is invited to try a number of avenues of approach, some of which, I admit, have no particularly satisfactory outcome but which would naturally be tested in any case. Indeed, I hope the alert and experienced reader will find rewarding byways that I have overlooked. Further, although this gymnasium is small, and perhaps ill equipped, my inventive target readers will already have at least a little experience in geology or related subjects and will soon find parts of many of their problems will yield to these elementary statistical methods.

B.1 Amphibole porphyroblasts in a muscovite schist

An exposure of this rock type is illuminated by low-angle sunlight. A linear structure is clearly visible, developed due to the preferred orientation of the muscovite crystals, whose mean grain size is about 0.5 mm. Because of the low-sun illumination, the amphibole porphyroblasts, up to 30 mm long, although scattered widely over the foliation surface of the schist, also *appear* to have a preferred orientation, parallel to that of the muscovite but much less strongly developed. The angles subtended with the muscovite lineation by 63 of these porphyroblasts are listed in Table B.1. The data are axial, thus modulo 180°. In order to test the *apparent* preferred alignment of these acicular amphibole crystals, first prepare a circular plot, then apply one or more of the tests of uniformity.

Table B.1 Orientations of the long axes of amphibole porphyroblasts on a foliation surface in muscovite schist, measured clockwise from a prominent lineation developed in the schistose matrix.

000	010	026	045	064	094	129	145	164
001	010	027	046	068	095	129	147	164
002	012	033	050	071	109	133	149	168
005	016	035	051	075	111	134	150	169
006	019	036	051	076	113	136	151	171
009	021	042	054	088	116	139	161	174
009	022	044	055	090	119	143	161	177

Table B.2 Directions of lines of bilateral symmetry of trilobites preserved on a bedding plane, measured clockwise from an arbitrary reference direction. The measured direction is taken as pointing towards the anterior of the individual.

002	021	061	120	188	205	227	340
007	023	074	151	192	214	236	353
014	028	087	163	194	215	299	354
021	047	098	165	196	216	337	

B.2 Orientation of trilobites

A bedding surface (actually illustrated on the front cover of Clarkson 1979) is crowded with the remains of trilobites. A few individuals are entire, rather more consist of thorax plus either cephalon or pygidium, but the majority is disaggregated. The initial impression is that the orientation of the part complete and complete individuals is completely irrational. However, this kind of impression is *always* worth checking, so verify, or otherwise, by using the data in Table B.2. Note that these orientations are *directions* with a sense towards the anterior of the individual and are thus modulo 360°. However, there may be some merit in also treating them as *axes* (thus modulo 180°) by a suitable transformation. Prepare circular plots, test for uniformity, and estimate von Mises parameters (mean direction, concentration: with confidence intervals) if appropriate. You may care to speculate on the palaeo-ecological significance of your results!

	M	*SSS*	*LL*	*Totals*
M				
SSS				
LL				
Totals				

Figure B.1 Contingency table showing the frequencies of occurrences of transitions from the rock type shown on the left, stratigraphically upwards and into the rock type shown at the top. Column totals show the number of transitions *to* that rock type, row totals show the number of transitions *from* that rock type.

B.3 Sedimentary successions I

Table B.3 shows an artificially generated stratigraphical succession that contains three rock types, sandstone, limestone and mudstone. Casual inspection may suggest that there is a perhaps moderately defined tendency for the beds to occur in the cyclic sequence mudstone–limestone–sandstone, then back to mudstone. However, there are many exceptions to this cycle, so a statistical assessment is needed.

Consider a contingency table of the form shown in Figure B.1. If the transitions occur at random, then the expected frequencies in any column of the table should be proportional to the respective row totals or, alternatively, the expected frequencies in any row should be proportional to the respective column totals. In either case, the expected frequency in any class is given by the column total times the row total divided by the overall sample size. Construct a contingency table for the data of Table B.3, calculate expected frequencies and apply the chi-squared test.

The probability of occurrence of a transition *from* one rock type *to* another (perhaps the same) is called a 'transition probability'. Such probabilities may be estimated by dividing the frequencies in the contingency table by their respective row totals. Unfortunately, the probability of occurrence of a transition *to* one rock type *from* another (perhaps the same) is also a transition probability, but 'backward-looking' rather than 'forward-looking'. These 'backward-looking' transition probabilities are estimated by dividing frequencies by their respective column totals.

Table B.3 An artificially generated stratigraphical succession that contains beds of mudstone (M), limestone (LL) and sandstone (SSS).

top	SSS	LL
SSS	LL	SSS
LL	M	LL
M	M	M
SSS	SSS	SSS
SSS	LL	LL
LL	M	M
M	LL	M
SSS	LL	SSS
M	M	LL
M	SSS	SSS
SSS	LL	LL
LL	LL	M
M	M	M
M	SSS	SSS
SSS	LL	SSS
LL	M	SSS
M	SSS	SSS
to next column	*to next column*	base

Table B.4 An artificially generated stratigraphical succession containing beds of mudstone (M), limestone (LL) and sandstone (SSS). The thickness of each bed is given (in meters).

top			
0.110 LL	0.120 LL	0.014 M	0.115 LL
0.668 SSS	0.528 SSS	0.532 SSS	0.693 SSS
0.100 LL	0.094 LL	0.424 SSS	0.528 SSS
0.019 M	0.014 M	0.113 LL	0.649 SSS
0.118 LL	0.256 SSS	0.459 SSS	0.105 LL
0.213 SSS	0.093 LL	0.018 M	0.367 SSS
0.536 SSS	0.132 LL	0.006 M	0.090 LL
0.014 M	0.137 LL	0.506 SSS	0.015 M
0.345 SSS	0.013 M	0.123 LL	0.507 SSS
0.107 LL	0.661 SSS	0.575 SSS	0.012 M
0.528 SSS	0.008 M	0.498 SSS	0.089 LL
0.544 SSS	0.493 SSS	0.122 LL	0.121 LL
0.076 LL	0.093 LLL	0.156 SSS	0.015 M
0.013 M	0.372 SSS	0.112 LL	0.007 M
0.510 SSS	0.519 SSS	0.605 SSS	0.634 SSS
0.104 LL	0.142 LL	0.102 LL	0.013 M
0.424 SSS	0.457 SSS	0.014 M	0.585 SSS
0.008 M	0.608 SSS	0.022 M	0.660 SSS
0.600 SSS	0.094 LL	0.505 SSS	0.015 M
0.295 SSS	0.014 M	0.017 M	0.110 LL
0.086 LL	0.015 M	0.012 M	0.437 SSS
0.016 M	0.121 LL	0.112 LL	0.313 SSS
0.211 SSS	0.093 LL	0.016 M	0.018 M
0.098 LL	0.172 SSS	0.470 SSS	0.368 SSS
0.481 SSS	0.095 LL	0.833 SSS	0.160 SSS
0.134 LL	0.582 SSS	0.016 M	0.110 LL
0.132 LL	0.011 M	0.077 LL	0.085 LL
0.014 M	0.094 LL	0.380 SSS	0.516 SSS
0.101 LL	0.013 M	0.016 M	0.013 M
0.638 SSS	0.742 SSS	0.015 M	0.807 SSS
0.127 LL	0.122 LL	0.074 LL	0.102 LL
0.419 SSS	0.013 M	0.017 M	0.013 M
0.395 SSS	0.066 SSS	0.121 LL	0.085 LL
0.356 SSS	0.096 LL	0.014 M	0.050 SSS
0.095 LL	0.365 SSS	0.109 LL	0.116 LL
0.428 SSS	0.078 LL	0.609 SSS	0.010 M
0.722 SSS	0.562 SS	0.018 M	0.014 M
			0.300 SSS
to next column	*to next column*	*to next column*	base

Bearing in mind the outcome of the chi-squared test, above, estimate these probabilities and decide whether they support a notion of cyclic sedimentation.

B.4 Sedimentary successions II

Table B.4 is a second artificially generated sedimentary succession, consisting of beds of mudstone, sandstone and limestone. Here, there may be no impression whatsoever of any regularity in the succession of beds, but these seemingly irregular sequences are always worth testing – the outcome means something! Using the methods of the last example, first apply the chi-squared test to the lowest 50 bed transitions. Then apply the same test to the highest 50 bed transitions and compare results.

Beds are also labelled according to their thickness. For each rock type, test these measurements for Gaussian distribution and, if appropriate, estimate mean thickness and accompanying confidence interval.

B.5 Dimensions and shapes of some fossils

The 'lengths' and 'heights' of individuals in two samples of *Gryphaea* are listed in Table B.5. Prepare scatter diagrams to show the distributions of

Table B.5 'Lengths' and 'heights' of individuals in two samples of the bivalve *Gryphaea* (dimensions in mm).

Length	*Height*	*Length*	*Height*	*Length*	*Height*
Sample 1					
91	86	78	74	85	79
75	66	63	52	71	69
56	49	68	54	85	75
36	34	60	62	87	86
84	80	53	45	101	97
36	32	55	52	84	85
52	49	55	50	42	49
48	48	58	49	95	85
53	39	41	37	64	50
39	37	54	43	21	21
53	46	42	45	34	34
38	29	39	35	109	108
48	39	23	16	87	78
36	30	22	16	62	52
35	31			80	71
36	33	Sample 2		72	66
68	62	98	91	34	33
		98	90	32	30
to next column		*to next column*			

these measurements; then devise a test or tests to compare the distribution of 'length' between the two samples, i.e. could individuals from one sample be larger, on the whole, than individuals from the other sample?

The 'shape' of an individual can be quantified by calculating the 'height' to 'length' ratio. Do this for all individuals in the two samples and use the runs test and the Mann–Whitney U test to test for differences between the populations from which the samples are drawn. Carefully review the results and comment upon them.

B.6 Garnet and biotite porphyroblasts

Using the notation of Chapter 4, the frequencies of occurrence of garnet and biotite porphyroblasts in the gneisses described there may be symbolised: $(BG) = 7$; $(Bg) = 26$; $(bG) = 26$ and $(bg) = 31$. Investigate the association of these porphyroblast species and in particular, compare the results obtained with critical regions of size 0.05 and of 0.01. What is the significance, if any, of the diagonal symmetry of the contingency table that arises?

B.7 Folded quartzite with lineation

A fold with limb dimensions approaching 1 km is developed in a metasedimentary quartzite formation in which sedimentation structures are preserved. These latter indicate that the stratigraphically older rocks occupy the core, making the fold an anticline. The constituent grains of the quartzite are elongate in a common orientation, thus imparting a visible linear fabric, or lineation, to hand specimens of the rock. The orientation of this linear fabric at a number of localities over the fold is presented in Table B.6, together with orientations of bedding at a somewhat larger number of localities.

Table B.6 The orientations of linear fabric and of bedding in a folded quartzite formation.

Angle of plunge/direction of plunge (clockwise from north) of the linear fabric:

32/325	38/337	45/316	36/300	32/293	50/304
42/290	30/269	60/254			

Angle of dip/direction of dip (clockwise from north) of bedding:

65/181	73/193	80/206	53/197	85/225	48/196
59/214	49/210	52/223	49/230	49/232	62/251
35/264	45/265	48/268	43/269	49/280	39/290
30/304	30/312	34/001	73/053	64/028	69/030
78/011					

Construct a stereographic projection of these axial data, representing bedding orientations by their downward-pointing normals. Use the methods of estimation associated with the Bingham distribution to find the mean direction of the lineation (i.e. vector $\mathbf{t}_3$) and the principal axes of the distribution of bedding normals. In connection with the latter, calculate, and construct on the stereographic projection, the 0.95 cone of confidence for the $\mathbf{t}_1$ vector, i.e. the best estimate for the fold axial direction. Is it realistic to suggest that the linear fabric is developed parallel to the hinge of the anticline?

Appendix C Numerical accuracy and errors

In the example in the section at the beginning of the book ('Evaluation of expressions'), suppose the calculator had been set to display the number of decimal places to its full capacity. Then:

$$d = 1.335802972 \text{ angstroms}$$

Some of these digits are surely spurious? The question is: 'how many?' The answer: 'all but two', as follows.

Decimal numbers that we substitute in formulae are rarely absolutely accurate because either they have been 'rounded off' or they are subject to experimental error. Consequently, as these inaccuracies pass through several stages of a calculation, they may accumulate to such an extent that the final answer is almost meaningless. The purpose of these notes is to show briefly how numbers may be rounded off, how the consequent inaccuracies may propagate, and how the accuracy of the final answer may be estimated (see Noble 1964, pp.1–11).

C.1 Decimal places and significant digits

The number of digits following the decimal point, including zeros, is known as the number of 'decimal places'. Thus, in presenting the chemical analysis of a rock, it is conventional to quote the concentrations of the major-element oxides to two decimal places, e.g. CaO = 12.08 wt%.

The number of 'significant figures' is the total number of digits in the answer that are supposed to be correct. Thus, the CaO concentration quoted as 12.08 wt% *may* be correct to four significant figures, but this depends on the analytical method used.

The number of significant figures in a result should always be made clear, because of ambiguities arising with non-significant zeros that are used simply to indicate the position of the decimal point. For example:

CaO = 12.08 wt% (four significant figures)

In the same rock, the concentration of a trace element would be quoted as, for example, Ba = 1200 p.p.m. (parts per million) correct, perhaps, to three significant figures. For example:

Ba = 1200 p.p.m. (three significant figures)
Ba = 0.00120 wt% (three significant figures)

Study these presentations carefully, especially the second: which are the significant and which are the non-significant zeros?

Ambiguities with zeros can be avoided by using the mantissa/exponent (or so-called) 'scientific' mode of expression. For example:

$$\text{Ba} = 1.20 \times 10^{3}\text{p.p.m.}$$
$$\text{Ba} = 1.20 \times 10^{-3}\text{wt\%}$$

The convention is that all figures given in the mantissa are significant.

C.2 Rounding off

Having decided on the number of significant figures or on the number of decimal places to be quoted, any digits in excess can be discarded by 'rounding off'. The rule is: to round off to n significant figures (or decimal places), retain the n significant digits and discard the others. If those discarded make less than half a unit in the nth place, leave the nth digit unchanged. If those discarded make over half a unit in the nth place, add a unit to the nth digit. If those discarded are exactly half a unit, round off to the nearest *even* digit. The object of the last is to avoid bias in that, on average, numbers will be 'rounded up' as frequently as 'rounded down'. For example, rounding off $d = 1.335802972$ to successively fewer decimal places produces the sequence:

1.3358
1.336
1.34
1.3

If your calculator can be set to round off automatically, see if it gives the same sequence.

C.3 Absolute errors: addition and subtraction

The absolute error in a number is the difference between the true value of that number and our approximation to that number. Thus we can never know the true value of the absolute error but we can estimate its maximum value, the maximum absolute error. For example, by quoting d = 1.34 (three significant figures), we are stating that the true value of d is somewhere between 1.344999...999 and 1.335000...001. Alternatively, we may say $d = 1.34 \pm 0.005$, where the maximum absolute error is 0.005.

When numbers are added or subtracted, the maximum absolute error in the result is the *sum* of the maximum absolute errors in the individual numbers.

Example. Assuming these numbers to be accurate in the significant figures quoted, determine their total:

	Number	*Maximum absolute error*
	15.9	0.05
	13.444	0.0005
	10.7246	0.00005
Totals	40.0686	0.05055

Therefore the result is really 40.1 with all digits significant. Note carefully that it is not permissible to round off before summing (try it and find out why).

C.4 Relative errors: multiplication and division

The relative error is the ratio of the absolute error to the true value of the number. Again, we cannot calculate it precisely but we can estimate the maximum relative error as the ratio of the maximum absolute error to our approximation of the number. For example, if $d = 1.34$ then the maximum relative error is about $0.005/1.34 = 0.0037$ (also subject to a smaller error that can be ignored).

In both multiplication and division, the maximum relative error of the result is the *sum* of the maximum relative errors of the component numbers.

Example. What is the value of d^3 when $d = 1.34$ (three significant figures)? The maximum relative error of d is 0.0037 (see above) so that the maximum relative error of d^3 is three times this: 0.0111. d^3 is $1.34 \times 1.34 \times 1.34 = 2.406104$. To estimate the maximum *absolute* error in d^3, multiply it by the maximum relative error:

$$2.406104 \times 0.0111 = 0.027 = 0.03 \text{ (approximately)}$$

Thus d^3 is subject to an error of ± 3 units in the second decimal place and is correct to only two significant figures:

$$d^3 = 2.4$$

C.5 Errors in functions

Formulae to be enumerated frequently contain functions such as log (x), sin (x), etc. The 'argument' of the function (the x) may be subject to error, so what is the corresponding error in the function? To a useful approximation, a change of δx in x produces a change of $f'(x) \times \delta x$ in $f(x)$ where $f(x)$ is the function and $f'(x)$ is its first derivative with respect to x. In the context of these notes, δx is the maximum absolute error in x. You will find lists of all the common first derivatives in any set of school mathematical tables that contains a section on the calculus.

Example. An angle x is measured as 35.2°, correct to all figures stated. What is $\sin(x)$ to its maximum number of significant figures?

The function $f(x) = \sin(x)$ so $f'(x) = \cos(x)$. Therefore the maximum absolute error of $\sin(x)$ is $\cos(x)$ times the maximum absolute error of x. In this example, $\sin(35.2°) = 0.576432315$ by an electronic calculator. The maximum absolute error of x is 0.05, so that the maximum absolute error of $\sin(x)$ is $\cos(35.2°)$ times 0.05, near enough to 0.04. Thus the function is subject to an error of ± 4 units in the second decimal place and has to be stated as:

$$\sin(35.2°) = 0.6$$

C.6 Conclusions

Errors arising from the limited precision of decimal numbers and the need to round off can be estimated easily, and the way in which these errors propagate through formulae can be analysed.

Example. In the Bragg-law formula mentioned in the first paragraph of this appendix, the top line (λ) has a maximum relative error of:

$$0.005/1.54 = 0.0032$$

The bottom line: $2 \times \sin(35.2°)$ has a maximum relative error of:

$$0.04/\sin(35.2°) = 0.071$$

The quotient thus has a maximum relative error approximately:

$$0.0032 + 0.071 = 0.074$$

and a maximum absolute error:

$$0.074 \times 1.335802972 = 0.099$$

Thus $d = 1.3 \pm 0.1$ as opposed to the electronic calculator answer of $d = 1.335802972$.

C.7 Rounding errors and computing

Different models of computers and programmable calculators differ greatly in the 'precision' with which they store numbers, and in the ways that they carry out their basic arithmetical operations. The principal limiting factor is the number of binary digits that constitutes the individual storage location, which constrains the number of significant decimal digits that are carried from one stage of a calculation to the next. In a long calculation, every intermediate result is rounded to the number of decimal digits characteristic of the machine, and so rounding errors may accumulate. In machines of limited precision, the number of decimal digits so carried may be nine or less. As a consequence, lengthy calculations, such as those of Equations 5.3a, b, c or of Table 9.4 may be prone to serious numerical instability. The careful programmer will always check the performance characteristics of the machine and analyse the likely behaviour of the expressions involved in the calculations before embarking on detailed programming.

Appendix D Notation

A area of a circular histogram
a size of critical region, probability of type I error, a regression constant
B presence of biotite
b absence of biotite, probability of type II error, a regression constant
C the number of classes in a classification
$\bar{C}$ mean of cosines
c a direction cosine, regression line intercept
$\mathbf{c}$ a unit vector
χ^2 (Greek 'chi') a test statistic
D Kolmogorov–Smirnov test statistic
d semi-apical angle of confidence sector or cone
d as a suffix, 'down' (see 'X')
E an expected frequency
e′ the base of natural logarithms, 2.718...(sometimes used as a suffix to 'log')
e as a suffix to denote direction, 'east' (see 'X')
F Fisher's variance-ratio statistic
F(x) the cumulative distribution function (cdf) of x
f frequency, class frequency
$f(x)$ the probability density function (pdf) of x
H probability of 'heads', a proportion or fraction in the range (O,1), statistic in the Kruskal–Wallis test, presence of hornblende
H_0 the null hypothesis
H_1 the alternative hypothesis
h absence of hornblende, 'heads'
i a suffix labelling individual specimens, classes or samples
$\mathbf{K}$ matrix of concentrations
k concentration parameter
$\mathbf{l}$ a unit vector in the mean direction
M population mean
$\bar{M}$ estimate of population mean
$\bar{M}_1, \bar{M}_2$ estimates of means from two samples
M_e median
m meridional angle, statistic in Hodges–Ajne test, regression line slope
N sample size
N_1, N_2 sample sizes in two-sample situations
n as a suffix, 'north' (see 'X')
O an observed frequency
π the ratio circumference : diameter of a circle, 3.14159...

R number of 'heads', sum of ranks
$\bar{R}$ mean resultant length
r radius of sector of circular histogram, number of runs, Pearson's coefficient
S number of samples or sub-samples
$\bar{S}$ mean of sines
s population standard deviation
$\bar{s}$ estimate of population standard deviation
Σ a summation symbol (see Eq. 1.2 or Eqs 4.2 & 4.3)
T probability of 'tails'
t 'tails', statistic of Student's t
$\mathbf{t}$ with suffix, principal axis of Bingham distribution
τ Kendall's rank correlation coefficient, principal value of Bingham distribution
U order statistic in Mann–Whitney and various circular tests
V_N statistic in Kuiper's test
ν number of degrees of freedom
w class width
X random variable; with suffixes n, e and d, angles made with reference axes pointing north, east, down
X_0 population mean direction
$\bar{X}_0$ estimate of mean direction
x deviate from the mean, a variable
z standardised deviate

Bibliography

Bingham, C. 1964. *Distributions on the sphere and on the projective plane.* PhD thesis, Yale University.

Chayes, F. 1971. *Ratio correlation.* Chicago: University of Chicago Press.

Clarkson, E. N. K. 1979. *Invertebrate palaeontology and evolution.* London: George Allen & Unwin.

Cochran, W. G. 1977. *Sampling techniques*, 5. New York: Wiley.

Davis, J. C. 1973. *Statistics and data analysis in geology.* New York: Wiley.

Frazer, R. A., W. J. Duncan and A. R. Collar 1963. *Elementary matrices*, 112. Cambridge: Cambridge University Press.

Hall, G. G. 1963. *Matrices and tensors.* London: Pergamon.

Hope, K. 1968. *Methods of multivariate analysis.* London: University of London Press.

Kamb, W. B. 1959. Ice petrofabric observations from the Blue glacier, Washington, in relation to theory and experiment. *J. Geophys. Res.* **64**, 1891–1909.

Koch, G. S. and R. F. Link 1970. *Statistical analysis of geological data*, vol. I. New York: Wiley.

Koch, G. S. and R. F. Link 1971. *Statistical analysis of geological data*, vol. II. New York: Wiley.

Krumbein, W. C. and F. A. Graybill 1965. *An introduction to statistical models in geology.* New York: McGraw-Hill.

Macbeath, A. M. 1964. *Elementary vector algebra.* London: Oxford University Press.

Mardia, K. V. 1972. *Statistics of directional data.* London: Academic Press.

Mann, H. B. and D. R. Whitney 1947. On a test of whether one of two random variables is stochastically larger than the other. *Ann. Math. Statist.* **18**, 50–60.

Meyer, S. L. 1975. *Data analysis for scientists and engineers*, 167–9, 342–9. New York: Wiley.

Noble, B. 1964. *Numerical methods 1*, 1–11. Edinburgh: Oliver & Boyd.

Phillips, F. C. 1954. *The use of the stereographic projection in geology.* London: Edward Arnold.

Siegel, S. 1956. *Nonparametric statistics for the behavioral sciences*, 6–17. New York: McGraw-Hill.

Sprent, P. 1977. *Statistics in action*, 32. London: Penguin.

Sprent, P. 1981. *Quick statistics, an introduction to non-parametric methods.* London: Penguin.

Stephens, M. A. 1970. Use of Kolmogorov–Smirnov, Cramer–von Mises and related statistics without extensive tables. *J. R. Statist. Soc.* **B32**, 115–22.

Till, R. 1974. *Statistical methods for the earth scientist.* London: Macmillan.

Wetherill, G. B. 1966. *Sequential methods in statistics.* London: Methuen.

Yule, G. U. and M. G. Kendall 1950. *An introduction to the theory of statistics*, Chs 1–3. London: Griffin.

Index

Bold entries refer to section numbers, and *italic* entries refer to text figures.